Ford Science

HANDBOOK OF FOOD ISOTHERMS:
WATER SORPTION PARAMETERS
FOR FOOD AND FOOD COMPONENTS

FOOD SCIENCE AND TECHNOLOGY

A SERIES OF MONOGRAPHS

Series Editors

A complete list of the books in this series appears at the end of the volume.

HANDBOOK OF FOOD ISOTHERMS: WATER SORPTION PARAMETERS FOR FOOD AND FOOD COMPONENTS

HÉCTOR A. IGLESIAS
JORGE CHIRIFE

Departamento de Industrias
Facultad de Ciencias Exactas y Naturales
Universidad de Buenos Aires, Argentina

1982

ACADEMIC PRESS
A Subsidiary of Harcourt Brace Jovanovich, Publishers

New York   London
Paris   San Diego   San Francisco   São Paulo   Sydney   Tokyo   Toronto

ACADEMIC PRESS, INC.
111 Fifth Avenue, New York, New York 10003

United Kingdom Edition published by
ACADEMIC PRESS, INC. (LONDON) LTD.
24/28 Oval Road, London NW1 7DX

Library of Congress Cataloging in Publication Data

Iglesias, Hector A.
  Handbook of food isotherms.

  (Food science and technology)
  Includes bibliographical references and index.
  1. Food--Water activity. I. Chirife, Jorge.
II. Title. III. Series.
TX553.W3134 1982   641.3   82-20735
ISBN 0-12-370380-8

PRINTED IN THE UNITED STATES OF AMERICA

82 83 84 85   9 8 7 6 5 4 3 2 1

To my son Diego Luis
from whom the time devoted to this book
has been withdrawn
H.A.I.

To my wife Margaret and daughter Paola
J.C.

CONTENTS

Preface   ix

Introduction   1

Graphical Representation of Experimental Data of Water
Sorption in Foods

   I.   Criteria Used for the Compilation and Representation
       of Data   8

  II.   Graphical Data   10

Mathematical Description of Isotherms

   I.   Introduction   262

  II.   Tables of Parameters for the Mathematical Description   264

 III.   Appendix: Nonlineal Regression Program Used for
       Determination of Parameters B(1) and B(2)   320

References   336

Products Index   344

PREFACE

This is the first English handbook entirely devoted to water vapor sorption data of foods and food components. The volume contains more than one-thousand isotherms with the mathematical description of over eight-hundred of these isotherms. A compilation of isotherms was published as a monograph in German in 1973 by W. Wolf, W. E. L. Spiess, and G. Jung ("Vasserdampf-sorptionsisotherm von Lebensmitteln," Fachgemeinschaft Allgemeine Lufttechnik im VDMA, Heft 18, Frankfurt) that included about four-hundred and sixty isotherms.

We shall be very obliged to readers who would call our attention to any aspect that has been neglected and to authors of papers that are not included.

Our first acknowledgment is to all the researchers who for many decades have been measuring food isotherms.

We also wish to thank the Secretaría de Estado de Ciencia y Tecnología de la República Argentina for its continuous support, and the University of Buenos Aires, Facultad de Ciencias Exactas y Naturales, for making this work possible.

The computational assistance of Dr. N. O. Lemcoff is also acknowledged. For the forbearance of our families, we are very grateful.

Héctor A. Iglesias
Jorge Chirife
Buenos Aires, 1981

INTRODUCTION

The water sorption isotherms of foods show the equilibrium relationship between the moisture content of foods and the water activity (a_w) at constant temperatures and pressures (Labuza, 1968; Loncin *et al.*, 1968; Gál, 1975). At equilibrium, the water activity is related to the relative humidity of the surrounding atmosphere by

$$a_w = \frac{p}{p_0} = \frac{\text{relative humidity } (\%)}{100}, \tag{1}$$

where p is the water vapor pressure exerted by the food material, p_0 the vapor pressure of pure water at temperature T_0, which is the equilibrium temperature of the system.

Water sorption isotherms are usually described as a plot of the amount of water sorbed as a function of the water activity, giving rise in most cases—but not in all—to curves of sigmoid shape.

At present, water activity determinations are a very common practice in food laboratories. This results from an increased recognition of the importance of water activity for characterizing the state of water in foods, namely, its availability for biological, physical, and chemical changes (van den Berg and Bruin, 1978; Rockland and Nishi, 1980).

CHANGES AFFECTED BY WATER ACTIVITY

We briefly summarize the main types of changes in foods affected by water activity.

A. Microbial Growth

Reduction of water activity provides a very important means of stabilizing food products. Most bacteria do not grow below $a_w = 0.90$ and most mold and yeast strains are inhibited between 0.88 and 0.80, although some osmophilic yeast strains can still grow down to 0.6. In a classic article Scott (1957) summarized the water relations of microorganisms; most of the principles suggested are still valid. Since then, many authors have reviewed the subject of water activity and microbial growth, among them Christian and

Waltho (1962), Pitt (1975), Leistner and Rödel (1975), Mossel (1975), and more recently Troller and Christian (1978).

B. Enzymatic Reactions

Enzymic reactions can occur in low-moisture foods when the enzymes have not been inactivated by heating. It has been shown that there is a correlation between the activity of enzymes and the water content of food. Although this correlation is complex, it is best expressed as a function of water activity rather than moisture content. The occurrence of enzymatic reactions in food at low water contents has been the subject of extensive studies over the past two decades (Acker, 1969; Multon and Guilbot, 1975; Potthast *et al.*, 1975, 1977a,b).

C. Nonenzymatic Browning and Lipid Oxidation

Nonenzymatic browning, which involves the reaction between carbonyl and amino compounds, and lipid oxidation are the major chemical deteriorative mechanisms that limit the stability of low- and intermediate-moisture foods.

The effects of water on these reactions are complex because water can act in one or more of the following roles: (a) as a solvent for reactants and for products, (b) as a product of reactions, and (c) as a modifier of the catalytic or inhibitory activities of other substances.

The influence of water activity and water content on the rate of nonenzymatic browning reactions in foods has been studied by a number of workers (Labuza *et al.*, 1970, Karel and Labuza, 1968; Eichner, 1975; Eichner and Ciner-Doruk, 1979; Resnik and Chirife, 1979; Rockland and Nishi, 1980). The effects of water activity on the kinetics of lipid oxidation have been reviewed by Labuza *et al.* (1969) and Labuza (1971, 1975).

D. Textural Changes

Textural changes in foods are also affected by water activity. Most work in this area has been focused on the toughening effect of freeze-drying and subsequent storage on meat and fish (Kapsalis *et al.*, 1971; Kapsalis, 1975).

E. Other Effects

Aroma retention in dried foods has been shown to depend greatly on water activity (Flink and Karel, 1972; Chirife and Karel, 1974). Water activity also affects important structural changes in foods such as amorphous–crystalline transformations in sugar-containing foods (Berlin *et al.*, 1968a; White and Cakebread, 1966; Iglesias *et al.*, 1975b).

A recent book by Troller and Christian (1978) adequately reviewed most effects of water activity on food stability, and the reader is referred to it for more detailed information.

F. Main Uses of the Sorption Isotherm

Besides predicting the microbial or physicochemical stability of foods, a knowledge of water sorption isotherms is also very important for engineering purposes related to concentration and dehydration.

The endpoint of any food dehydration is generally determined by the desired water activity of the finished product. In dried foods at moisture contents where the equilibrium relative humidity is below saturation, a likely mechanism of moisture transport is vapor phase diffusion (King, 1968). In these conditions, the effective diffusion coefficient of water D_{eff} is related to various physical properties of the food as well as environmental conditions and depends on an isotherm factor $(\partial a_w / \partial x)_T$. This factor relates the water activity in the food to the moisture content (Viollaz et al., 1978) and is the inverse of the slope of the moisture sorption isotherm.

One use of isotherms at two or more temperatures is in predicting sorption values at other temperatures. Food isotherms at several temperatures usually show a decrease in the amount sorbed with an increase in temperature at constant water activity (Labuza, 1968; Iglesias and Chirife, 1976b; Bandyopadhyay et al., 1980). This means that these foods become less hygroscopic with an increase of temperatures. From the well-known thermodynamic relationship,

$$\Delta F = \Delta H - T \, \Delta S$$

as $\Delta F < 0$ (sorption is a spontaneous process) and $\Delta S < 0$ (the sorbed molecule has less freedom),

$$\Delta H < 0.$$

Then, an increase of temperature does not favor water sorption.

However, it is known that some sugars (or foods containing sugars) show an opposite trend in their isotherms; that is, they become more hygroscopic at higher temperatures because of the dissolving of sugar in water (Iglesias et al., 1975a; Audu et al., 1978; Bandyopadhyay et al., 1980). Usually sorption phenomena in foods obey the Clausius–Clapeyron relationship (Saravacos and Stinchfield, 1965; Iglesias and Chirife, 1976b; Mazza and Le Maguer, 1978). The temperature dependence of the isotherm may be expressed as

$$\left(\frac{\partial \ln a_w}{\partial T} \right)_{x_a} = \frac{Q^{st}_{x_a}}{RT^2}, \tag{2}$$

where $Q^{st}_{x_a}$ is the net isosteric heat of sorption, which is a differential molar quantity, and x_a the amount of water sorbed.

The net isosteric heat of sorption may be calculated from Eq. (2) by plotting the sorption isostere as ln a_w vs. $1/T$ and determining the slope, which equals $-Q_x^{st}/R$. By this method it is not necessary to assume that Q_x^{st} is invariant with temperature, but the application of the method requires the measurement of sorption isotherms at more than two temperatures.

The isosteric heat may be also calculated from the integrated form of Eq. (2) applied to sorption isotherms measured at two temperatures,

$$Q_{x_a}^{st} = R\left(\frac{T_1 T_2}{T_2 - T_1}\right)\left(\ln \frac{a_{w_2}}{a_{w_1}}\right). \tag{3}$$

Knowledge of the net heat of sorption provides an indication of the binding energy of water molecules and has some bearing on the energy balance of drying operations.

The graphical determination of the isosteric heat may be carried out by carefully plotting the tabulated sorption data with a scale large enough for preserving the precision of the data; at values of moisture content taken at convenient intervals, the isosteric water activities are read and the values used with Eq. (2) or (3). In this way it is possible to predict the water activity at some unknown temperature in the range covered by the known sorption isotherms. Because of some irreversible changes in food materials subjected to high temperatures, predictions should be limited to temperature ranges not very far from the available one; otherwise deviations may occur (Bandyopadhyay et al., 1980).

In some cases the temperature dependence of water sorption isotherms can be estimated using isotherm equations containing additional constants characteristic for the food material (Chen and Clayton, 1971; Iglesias and Chirife, 1976d).

In analyzing sorption behavior of complex food mixtures, it is often desirable to be able to calculate the sorption isotherm of the mixture. In some cases this is possible from knowledge of component isotherms and weight fractions of components (Iglesias et al., 1980). In formulating multicomponent dehydrated foods, such as dried soups and sauces, the sorption isotherm of each component must be known in order to predict and to avoid an undesirable transfer of water from substances of high water activity to those of lower a_w. Salwain and Slawson (1959) developed a procedure to predict moisture transfer in combinations of dehydrated foods, from the knowledge of the sorption isotherms of the individual components. In this procedure, which is based on the idea that at equilibrium all products have the same water activity, portions of the isotherm of each component are approximated by straight lines and slopes evaluated in order to calculate the equilibrium a_w. The water activity of the mixture is given by

$$(a_w)_M = \frac{a_{w_1} S_1 m_1 + a_{w_2} S_2 m_2 + \cdots + a_w S_n m_n}{S_1 m_1 + S_2 m_2 + \cdots + S_n m_n}, \tag{4}$$

where $(a_w)_M$ is the water activity of the mixture at equilibrium;

$a_{w_1}, \ldots, a_w$ are initial water activities of n components; $S_1, \ldots, S_n$ are isotherm slopes of components; and $m_1, \ldots, m_n$ are weights of dry solids of components. A limitation of this method is the assumption of a linear isotherm.

Iglesias *et al.* (1979) developed a computer technique based on the classical BET isotherm equation (Labuza, 1968) to predict equilibrium conditions after mixing dehydrated foods. This technique is applicable to mixtures of dried foods in the range $0.05 \leqslant a_w \leqslant 0.40$, which is the range of more practical interest when concerning mixtures of dehydrated foods, e.g., dried soup or sauce mixtures.

Packaging dehydrated foods is one of the most important application of sorption isotherms, since the prediction of storage life of dehydrated foods packaged in flexible films is of obvious importance in the area of food preservation. In the last few years there has been an increased interest in the development of mathematical models for optimization of flexible film packaging of dehydrated foods (Mizrahi *et al.*, 1970; Quast *et al.*, 1972; Labuza *et al.*, 1972; Iglesias *et al.*, 1977). Development of those models has allowed the prediction of storage stability as a function of package properties for dried foods deteriorating through different moisture-sensitive reactions. They were based on the combination of kinetic data for the deteriorative reactions, the water sorption isotherm of the food, and the permeability characteristics of the package. The rate of transport of water vapor through a flexible film is given by

$$\frac{dw}{d\theta} = \frac{PA}{e} (p_e - p), \tag{5}$$

where w is the weight of water transferred across the film, θ the time, P the permeability of the film, e the film thickness, A the area of the film, p_e the vapor pressure of water outside the film, and p the vapor pressure of water on the other side of film.

In order to solve for water gain it is usually assumed (Mizrahi *et al.*, 1970; Iglesias *et al.*, 1977) that the water entering the package rapidly equilibrates with the food. The internal vapor pressure p is therefore solely determined by the water sorption isotherm of the food.

In order to solve for water gain the isotherm equation may be rearranged and substituted into Eq. (5). In the case of a linear isotherm the result is directly integratable; if not the equation may be numerically evaluated by using computational techniques (Iglesias *et al.*, 1977).

G. Influence of Pretreatments, Chemical Composition, and Species Differences in Regard to Sorption Isotherms

The history and the pretreatment of the food sample may have a significant influence on its water sorption isotherm. These influences are not very well known and have hardly been studied (van den Berg and Leniger, 1976). For this reason, and as noted by van den Berg and Leniger (1976), "one cannot speak strictly about the sorption isotherm as a well defined physical property, without a clear statement of how the results have been obtained. Both the history and the pretreatment immediately before the measurement of the equilibria in practice can have a significant influence on the position of the curve."

Saravacos (1967) reported that the method of drying (air-, puff-, or freeze-drying) affects to some extent the water adsorption isotherms of apple and potatoes; the freeze-dried products sorbed more water than the puff- and air-dried materials. Iglesias and Chirife (1976b) determined the water adsorption isotherms of precooked beef previously dried with air at different temperatures and found that the higher the drying temperature the lower the sorption. San José et al. (1977) reported that adsorption isotherms of freeze- and spray-dried lactose-hydrolyzed milk were identical, but desorption curves were influenced by the drying method. Hayakawa et al. (1978) also reported that different drying methods (spray- and freeze-drying) significantly influenced moisture adsorption isotherms of dried coffee products. Lewicki and Lenart (1975) studied the effect of different drying procedures on adsorption isotherms of carrots. Bolin (1980) reported that isotherms of sun-dried raisins were slightly lower than those of the vacuum-dried grapes.

Heldman et al. (1965) reported that preheat treatment influenced the adsorption isotherms of nonfat dry milk. Greig (1979) showed that denaturation of native cottage cheese whey protein had no effect on the ability of the powder to sorb water at low relative humidities, but water sorption at high relative humidities was significantly reduced.

Lewicki and Lenart (1975) determined the adsorption isotherms of carrots as influenced by preliminary treatments (blanching, freezing) before drying. The effect of rate of freezing of the adsorption isotherms of freeze-dried muscle fibers has been reported by Mac Kenzie and Luyet (1967). Hayakawa et al. (1978) found that decaffeination of soluble coffee affected moisture sorption by spray-dried coffee; the decaffeinated sample adsorbed greater amounts of moisture in comparison with the amount adsorbed with the undecaffeinated sample. However, there was an entirely opposite influence of decaffeination on moisture adsorption with roast and ground samples of coffee.

Particle-size distribution of the food usually does not influence the sorption isotherm. Gur-Arieh et al. (1967) showed that the water adsorption isotherm of wheat flour was independent of particle-size distribution. King et al. (1967) similarly reported that adsorption and desorption isotherms of freeze-dried turkey were not affected by pore size of the dried food.

Moisture sorption isotherms of foods represent the integrated hygroscopic properties of its individual components. Consequently any modification in chemical composition due to a chemical change, as a consequence of varietal differences or degree of maturity, may in turn influence the sorption isotherm. For example, Malthouthi *et al.* (1981) reported that the degree of proteolysis of gruyere cheese has an important effect on its water sorption behavior. Komeyasu and Iyama (1974) found that the equilibrium moisture contents of spray-dried citrus juice were affected by its pulp content. Bolin (1980) observed that maturity (which affects sugar content) of fresh raisins influenced the sorption isotherms. It is well known that the presence of fat modifies the water sorption capacity of foods; i.e., the higher the fat content the lower the equilibrium moisture content for a specific water activity (Saravacos, 1969; van Twisk, 1969; Heldman *et al.*, 1965). For this reason and in order to allow a comparison between similar foods but with different fat contents, moisture content should be given on a percentage nonfat dry basis, grams of water per 100 grams of nonfat dry material. This procedure is based on the assumption that the fat does not sorb water (or at least very little as compared to other food components). Iglesias and Chirife (1977a) have shown that adsorption isotherms of minced beef samples formulated with varying proportions of fat were coincident when expressed on a fat-free basis.

The influence of variety on moisture sorption isotherms of several different foods has been studied by a number of workers. They include— among many others—Pixton and Warburton (1977), who compared adsorption and desorption isotherms of three different varieties of rapeseed; Juliano (1964), who compared water sorption data by different varieties of rough rice; and Hubbard *et al.* (1957), who did the same for various varieties of wheat and corn. Strolle and Cording (1965) determined the adsorption isotherms of potato flakes made from eight potato varieties but found that regardless of variety and also geographical origin, the flakes had similar isotherms. Putranon *et al.* (1979) also reported sorption isotherms of two different cultivars of paddy rice and found that the effect of cultivar was small.

H. Recommendations for the Use of Isotherms

This handbook attempts to present in a practical way a compilation from the literature of the experimental data on water sorption isotherms of foods and food components. The data are presented in a graphical way and, whenever possible, through an isotherm equation in order to provide its mathematical description.

For the reasons discussed above, it is clear that there is not a single isotherm for a given product; pretreatments, maturity, variety, and chemical changes may all somehow influence the shape of the isotherm. For this reason the reader should select the sorption data that most closely resemble his particular interest. Even so, some fluctuations may be expected in the data due to the inherent variability of food materials.

GRAPHICAL REPRESENTATION OF EXPERIMENTAL DATA OF WATER SORPTION IN FOODS

I. CRITERIA USED FOR THE COMPILATION AND REPRESENTATION OF DATA

A. Source of Data

Water sorption data were obtained from a variety of available sources, most of them scientific and technical journals, papers presented at meetings, faculty and institute publications and reports.

B. Presentation of Data

Data are presented as plots of moisture content vs water activity. In the majority of cases, moisture content is given on a percentage dry basis (X), i.e., grams of water per 100 g of dry material. In some cases, however, a percentage wet basis (W) is used, i.e., grams of water per 100 g of total basis, and this is indicated. The relationship between these two moisture bases is

$$X = W/(1 - W). \qquad (6)$$

For some fat-containing foods, moisture content is given on a percentage nonfat dry basis, i.e., grams of water per 100 g of nonfat dry material.

Every illustration of sorption isotherms is accompanied by a summary of available information regarding the type and origin, chemical composition, physical nature, and history of the sample, together with a brief indication of the method of determination. In addition to this, each isotherm needs specification of the temperature of the experiment and direction of sorption, i.e., ad- or desorption. If this information is not given it is because it is not available in the original data source.

It is important to note that it is the usual practice in the sorption literature to draw the isotherm curves by graphical interpolation of experimental data points, these not being included in the plot. The reason for this procedure is usually to avoid overlapping of data when plotting more than one isotherm for each illustration. In other cases, authors give the data in tabulated form, but they are not the "experimental" ones but "round-off" values at specified water activities obtained after graphical interpolation.

For these reasons, and for the sake of uniformity and clarity of presentation, it was decided to reproduce only the isotherm curves drawn by the original author; when data were given in tabulated form curves were drawn by graphical interpolation.

C. Methods of Determination of Water Sorption Isotherms

In order to understand adequately the indication regarding the method employed, a brief review is given here. Procedures for obtaining water sorption isotherms in foods have been described in detail by Taylor (1961), Troller and Christian (1978), and more specifically by Gál (1975), to whom the reader is referred for more details. The principal methods are gravimetric, manometric, and hygrometric.

1. Gravimetric

Methods with continuous registration of weight changes. In this case a balance (recording electrobalance or a quartz spring balance) is a fixed part of the apparatus and weight changes can be determined continuously. This method is usually carried out in an evacuated system to accelerate diffusion of water molecules from the reservoir to the sample.

Dynamic systems. In these cases circulated air is the carrier for the transfer of water vapor to and from the sample. Precise weighings are possible at a constant flow rate of air around the sample.

Methods with discontinuous registration of weight changes.

Static systems. The most common one is that in which the material is placed in vacuum desiccators (or a closed jar) containing saturated salt solutions or sulfuric acid solutions which give a certain equilibrium relative humidity. Salt or sulfuric acid solutions for the various relative humidities are available (Labuza *et al.*, 1976; Troller and Christian, 1978). A vacuum may be created to accelerate equilibrium.

Dynamic systems. An air stream of known relative humidity is forced to pass over the sample.

2. Manometric

The vapor pressure of water in equilibrium with a food at a given moisture content is measured by a sensitive manometric device.

3. Direct Hygrometer Methods

The equilibrium relative humidity of a small amount of air in contact with a food at a given moisture content is measured by a hygrometer device. Dew-point or electric hygrometers, are frequently used.

D. Definition of the Terms Sorption, Adsorption, and Desorption Used in the Text

In the past there has been a great deal of use of the term sorption for adsorption; this is to indicate the direction of the sorption process (sorption or desorption). However, for reasons discussed below, we shall use the term *adsorption* to indicate when the isotherm is made by placing a dry material into atmospheres of increasing relative humidities and by measuring the gain in weight. The *desorption* isotherm will indicate that it is obtained by placing the initially wet material under the same relative humidities but by measuring the loss in weight.

In some of the literature, the isotherms cannot be considered "totally" adsorption or desorption. This occurs when the initial moisture is not low enough to lead to water gain at all relative humidities; then some points of the isotherm are obtained from the desorption one. In this case we use the term *sorption* isotherm.

II. GRAPHICAL DATA

See Figs. 1–501.

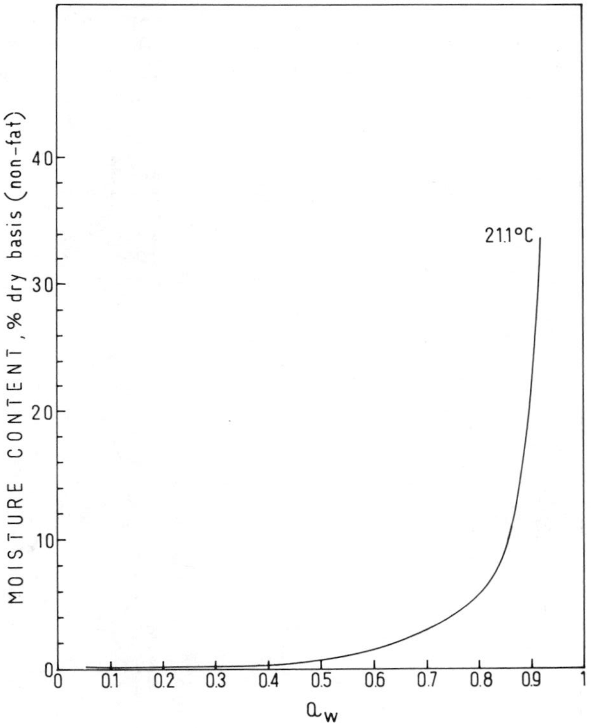

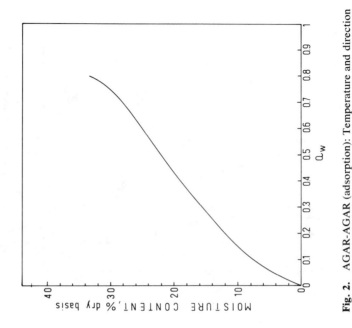

Fig. 1. Actomyosin, beef (adsorption, 21.1°C): Extracted from longissimus dorsi muscle of beef and freeze-dried at 26.7°C plate temperature. Method: electrobalance assembly. (Palnitkar and Heldman, 1971.)

Fig. 2. AGAR-AGAR (adsorption): Temperature and direction of sorption are not specified (Duckworth, 1972).

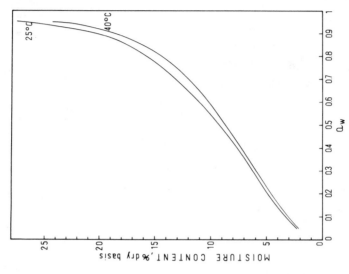

Fig. 4. Albumin, egg, coagulated (adsorption): Obtained from fresh hen's eggs; recrystallized three times and dialyzed against distilled water; coagulated by heating in boiling water bath and freeze-dried. Method: Static-desiccator (sulfuric acid solutions). (Bull, 1944.)

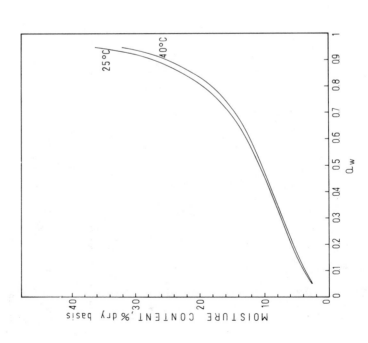

Fig. 3. Albumin, egg (adsorption): Obtained from fresh hen's eggs; recrystallized three times, dialyzed against distilled water, and air-dried at room temperature. Method: Static-desiccator (sulfuric acid solutions) (Bull, 1944).

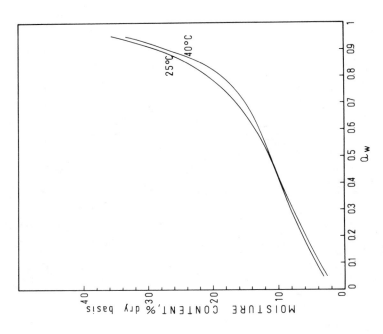

Fig. 6. Albumin, serum (adsorption): Prepared from healthy horse blood. Method: Static-desiccator (sulfuric acid solutions). (Bull, 1944.)

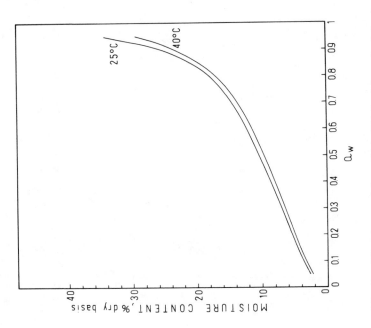

Fig. 5. Albumin, egg (adsorption): Obtained from fresh hen's eggs; recrystallized three times, dialyzed against distilled water, and freeze-dried. Method: Static-desiccator (sulfuric acid solutions). (Bull, 1944.)

13

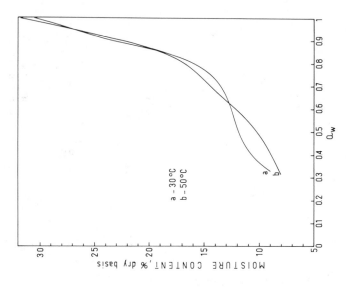

Fig. 7. Alginic acid (adsorption): Method: Dynamic (air of known relative humidity circulating over the samples). (Shotton and Harb, 1965.)

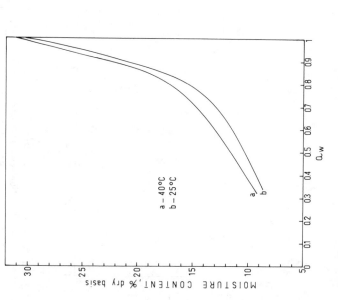

Fig. 8. Alginic acid (adsorption): Method: Dynamic (air of known relative humidity circulating over the samples). (Shotton and Harb, 1965.)

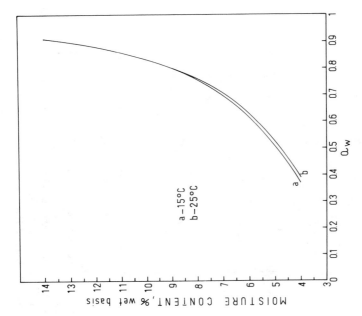

Fig. 9. Almonds, Californian (adsorption): Californian sweet-shelled, 1974 crop, received with 5.3% moisture content (wet basis) and 56.2% (dry basis) oil content; samples dried at 30°C to bring $a_w < 0.30$. (a) 15, (b) 25, (c) 35°C. Method: Dew-point. (Pixton and Henderson, 1979.

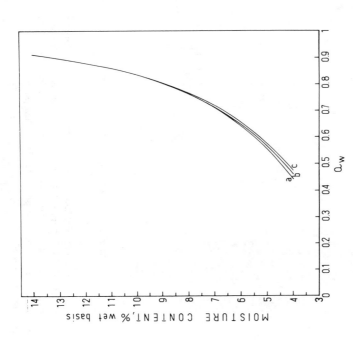

Fig. 10. Almonds, Californian (desorption): Californian sweet-shelled, 1974 crop, received with 5.3% (wet basis) moisture content and 56.2% (dry basis) oil content; a_w was raised to 0.95 by adding water before desorption. Method: Dew-point. (Pixton and Henderson, 1979.)

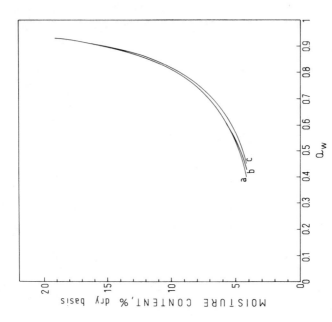

Fig. 12. Almonds, Moroccan bitter (adsorption and desorption): Moroccan bitter, 1976 crop, received with 4.9% (wet basis) moisture content and 54.2% (dry basis) oil content; samples for adsorption dried at 30°C to bring $a_w < 0.30$; for desorption water was added to raise a_w to 0.95; (a) 15, (b) 25, (c) 35°C. Method: Dew-point. (Pixton and Henderson, 1979.)

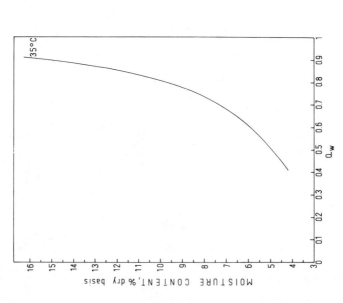

Fig. 11. Almonds, Californian (desorption, 35°C): Californian sweet-shelled, 1974 crop, received with 5.3% (wet basis) moisture content and 56.2% (dry basis) oil content; a_w was raised to 0.95 by adding water before desorption. Method: Dew-point. (Pixton and Henderson, (1979.)

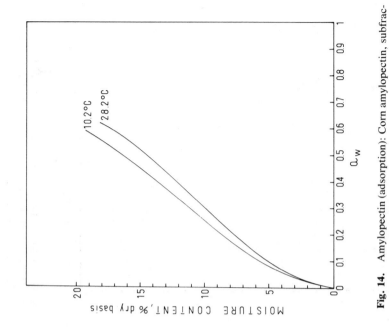

Fig. 14. Amylopectin (adsorption): Corn amylopectin, subfraction III, supplied by Professor W. Hassid. (Volman et al., 1960.)

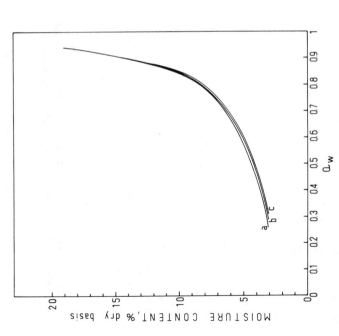

Fig. 13. Almonds, Moroccan sweet (adsorption and desorption): Moroccan sweet, 1976 crop, received at 4.5% (wet basis) moisture content and 60.2% (dry basis) oil content; samples for adsorption dried at 30°C to bring $a_w < 0.30$; for desorption water was added to raise a_w to 0.95; (a) 15, (b) 25, (c) 35°C. Method: Dewpoint. (Pixton and Henderson, 1979.)

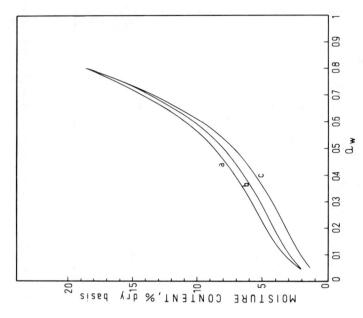

Fig. 15. Amylose (adsorption): Corn amylose subfraction, Schoch number C-148/150-A, 14 a; recrystallized from *n*-butanol. (Volman *et al.*, 1960.)

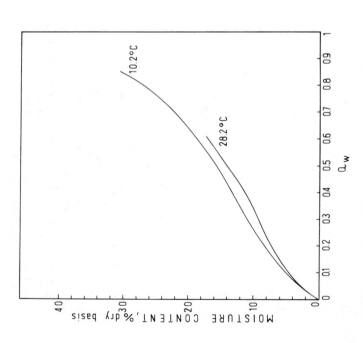

Fig. 16. Anise (adsorption): Vacuum-dried at 30°C before adsorption; (a) 5, (b) 25, (c) 45°C. Method: Jar with air agitation (sulfuric acid solutions). (Wolf *et al.*, 1973.)

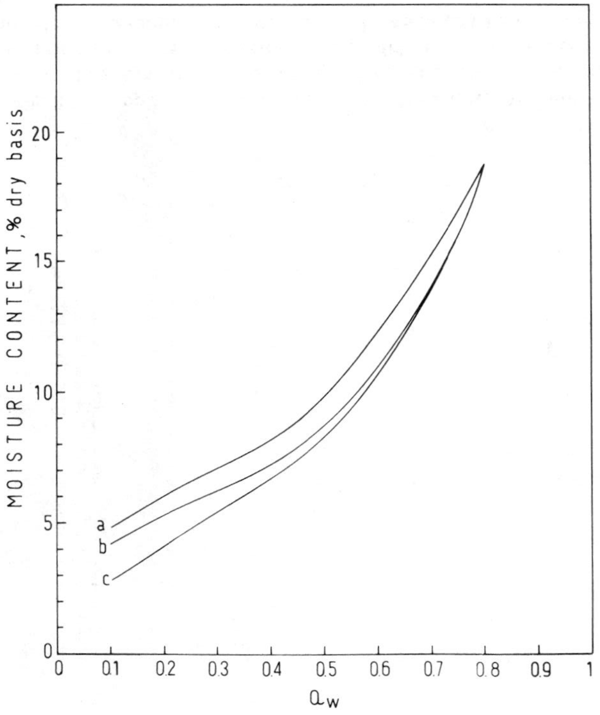

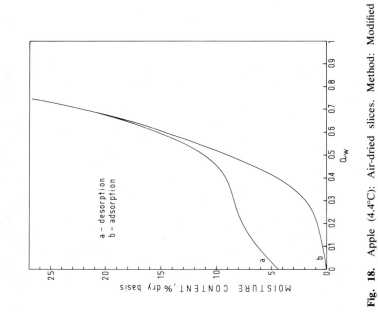

Fig. 17. Anise (desorption): Vacuum-dried at 30°C before adsorption; desorption following humidification at 90% relative humidity: (a) 5, (b) 25, (c) 45°C. Method: Jar with air agitation (sulfuric acid solutions). (Wolf et al., 1973.)

Fig. 18. Apple (4.4°C): Air-dried slices. Method: Modified McBain–Bakr vacuum sorption apparatus. (Wolf et al., 1972.)

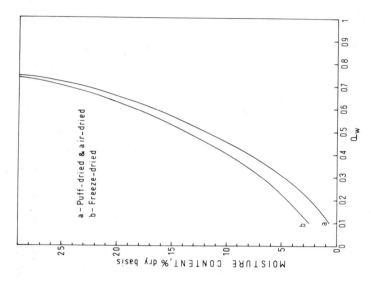

Fig. 20. Apple (adsorption, 30°C): Variety McIntosh cut into disk 15 mm in diameter and 1 mm thick and freeze-dried at room temperature or air-dried at 71.1°C. Method: Vacuum sorption apparatus with quartz-spring balance. (Saravacos, 1967.)

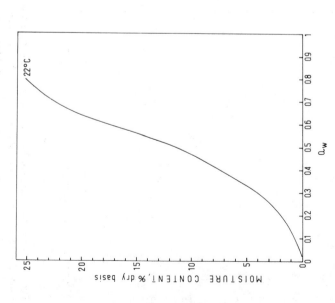

Fig. 19. Apple (adsorption, 22°C): Apple (variety Antonowka) cubes at about 0.8 cm were air-dried at 80°C and then in vacuum at 55°C prior to isotherm determination. Method: Static-desiccator (saturated salt/sulfuric acid solutions). (Lewicki and Lenart, 1977.)

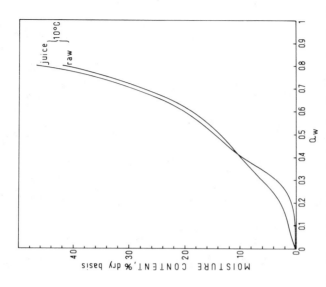

Fig. 22. Apple (adsorption): Variety Bramley Seedling; freeze dried; soluble constituents 85.1% (dry basis). Method: Static-jar (sulfuric acid solutions). (Gane, 1950.)

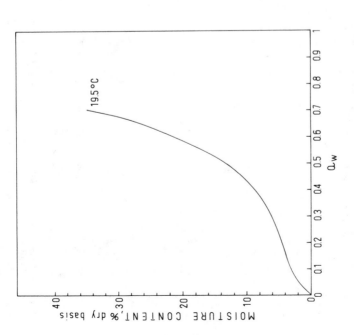

Fig. 21. Apple (desorption, 19.5°C): A freeze-dried sample was ground and allowed to equilibrate to a moisture content of about 20% (dry basis) before isotherm determination. Method: Manometric apparatus. (Taylor, 1961.)

21

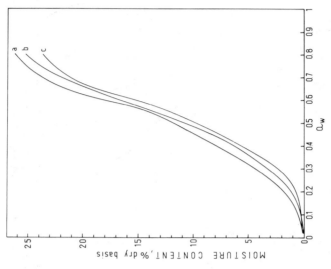

Fig. 24. Apple (osmotically treated, adsorption 22°C): Variety Antonowka; samples subjected to osmotic dehydration in sucrose solutions, then air-dried at 80°C, followed by vacuum-drying at 55°C before adsorption; 1-hr osmosis at (a) 20, (b) 30, (c) 40°C. Method: Static-desiccator (saturated salt/sulfuric acid solutions). (Lewicki and Lenart, 1977.)

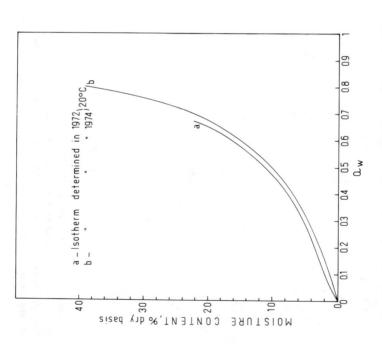

Fig. 23. Apple juice (adsorption, 20°C): Spray-dried apple juice before (1972) and after (1974) two-year storage. Method: Static-desiccator (saturated salt/sulfuric acid solutions). (Lewicki, 1976.)

22

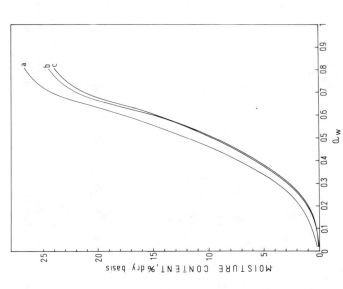

Fig. 25. Apple (osmotically treated, adsorption 22°C): Variety Antonowka; samples subjected to osmotic dehydration in sucrose solutions, then air-dried at 80°C, followed by vacuum-drying at 55°C before adsorption; 3.5-hr osmosis at (a) 20, (b) 40, (c) 30°C. Method: Static-desiccator (saturated salt/sulfuric acid solutions). (Lewicki and Lenart, 1977.)

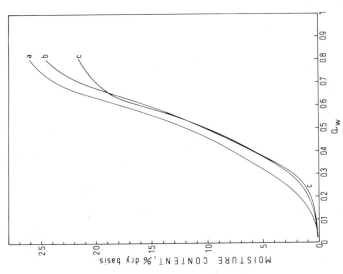

Fig. 26. Apple (osmotically treated, adsorption 22°C): Variety Antonowka; samples subjected to osmotic dehydration in sucrose solutions, then air-dried at 80°C, followed by vacuum-drying at 55°C before adsorption; 6-hr osmosis at (a) 20, (b) 30, (c) 40°C. Method: Static-desiccator (saturated salt/sulfuric acid solutions). (Lewicki and Lenart, 1977.)

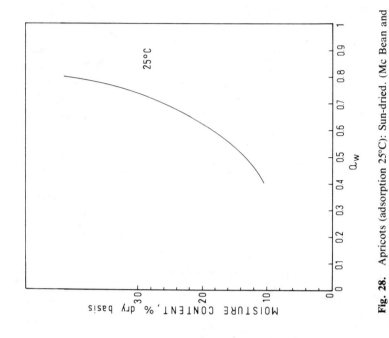

Fig. 28. Apricots (adsorption 25°C): Sun-dried. (Mc Bean and Wallace, 1967.)

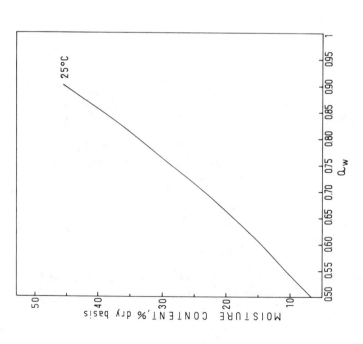

Fig. 27. Apricots (adsorption, 25°C). Fresh ripe apricots, variety Raanana; air-dried at 55°C before adsorption. Method: Static-desiccator (saturated salt solutions). (Harel *et al.*, 1978.)

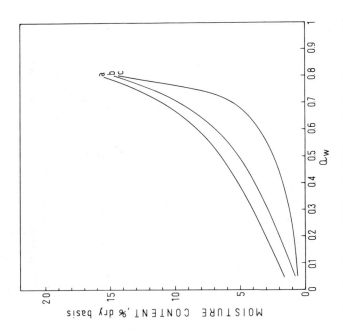

Fig. 30. Avocado (adsorption): Freeze-dried and vacuum-dried at 30°C before adsorption; (a) 25, (b) 45, (c) 60°C. Method: Jar with air agitation (sulfuric acid solutions). (Wolf *et al.*, 1973.)

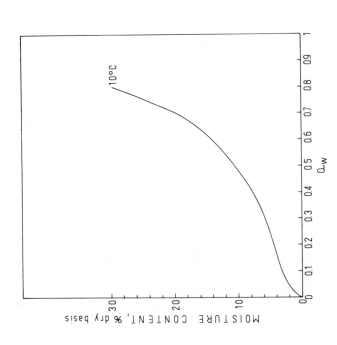

Fig. 29. Asparagus (adsorption, 10°C) Scalded, freeze-dried; soluble constituents 28.8% (dry basis). Method: Static-jar (sulfuric acid solutions). (Gane, 1950.)

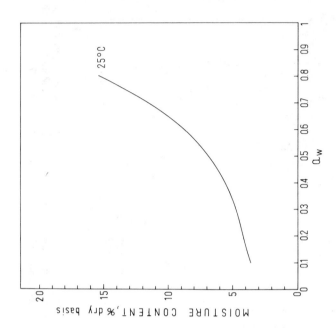

Fig. 32. Avocado (desorption, 25°C): Freeze-dried and vacuum-dried at 30°C before adsorption; desorption following humidification at 90% relative humidity. Method: Jar with air agitation (sulfuric acid solutions). (Wolf et al., 1973.)

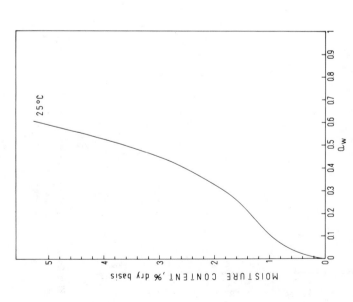

Fig. 31. Avocado (adsorption, 25°C): Variety Hass; freeze-dried at 40°C plate temperature. Method: Static-desiccator (sulfuric acid solutions). (Lladser and Piñaga, 1975.)

26

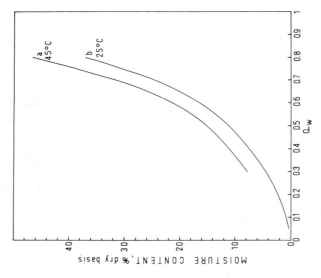

Fig. 34. Banana (adsorption, 25°C; desorption, 45°C). (a) Freeze-dried and vacuum-dried at 30°C before adsorption; desorption following humidification at 90% relative humidity; (b) air-dried; adsorption. Method: Jar with air agitation (sulfuric acid solutions). (Wolf *et al.*, 1973.)

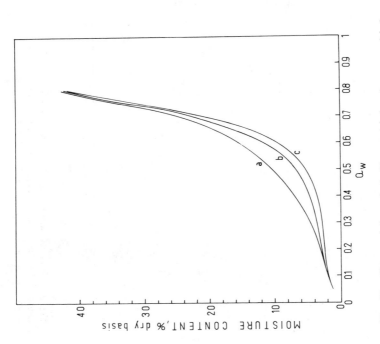

Fig. 33. Banana (adsorption): Freeze-dried and vacuum-dried at 30°C before adsorption; (a) 25, (b) 45, (c) 60°C. Method: Jar with air agitation (sulfuric acid solutions). (Wolf *et al.*, 1973.)

27

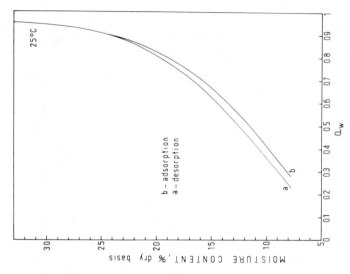

Fig. 36. Barley (adsorption and desorption, 25°C): Freshly harvested barley dried at 6% (wet basis) moisture content with air at 30°C and used for the adsorbing samples; desorbing samples prepared by addition of water to raise moisture content to about 25% (wet basis). Method: Dew-point. (Warburton and Pixton, 1973.)

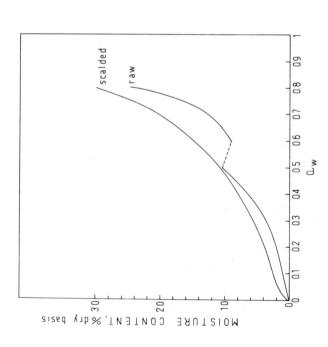

Fig. 35. Banana (adsorption): Gros Michael Ripe; soluble constituents 55.7% (dry basis); freeze-dried; 10°C. Method: Static-jar (sulfuric acid solutions). (Gane, 1950.)

28

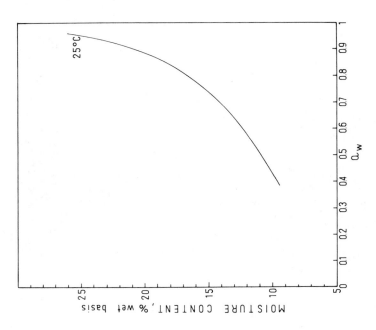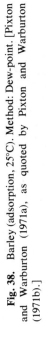

Fig. 38. Barley (adsorption, 25°C). Method: Dew-point. [Pixton and Warburton (1971a), as quoted by Pixton and Warburton (1971b).]

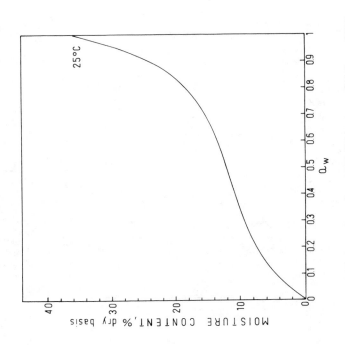

Fig. 37. Barley (adsorption, 25°C) [Coleman and Fellows (1925) as quoted by Chen (1971).]

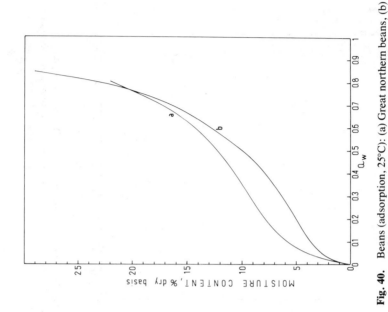

Fig. 40. Beans (adsorption, 25°C): (a) Great northern beans, (b) freeze-dried samples. Method: (b) Static-desiccator (saturated salt solutions). [(a) Weston and Morris (1954), as quoted by Rockland (1957); (b) Lafuente and Piñaga (1966).]

Fig. 39. Beans (adsorption): (a) Variety "rosinha" (Brazil), 25°C; (b) white pea beans, 25°C; (c) dry pinto beans, 21 ± 3°C. Method: (c) Vapor pressure manometer. [(a) Jordão and Stolf (1969–70), as quoted by McCurdy et al. (1980). (b) Dexter et al. (1955), as quoted by McCurdy et al. (1980). (c) McCurdy et al. (1980).]

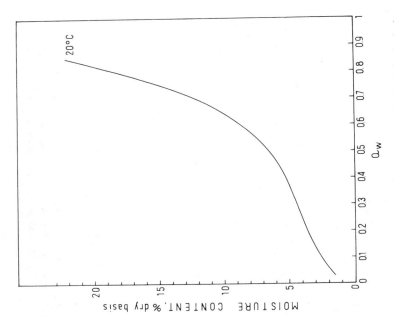

Fig. 42. Beef, dried (adsorption, 20°C): Method: Static-desiccator (saturated salt/sulfuric acid solutions). (Lewicki and Brzozowski, 1973.)

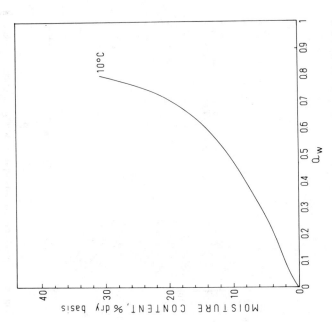

Fig. 41. Beans, runner (adsorption, 10°C): Scalded and then freeze-dried; soluble constituents: 42.4% (dry basis). Method: Static-jar (sulfuric acid solutions). (Gane, 1950.)

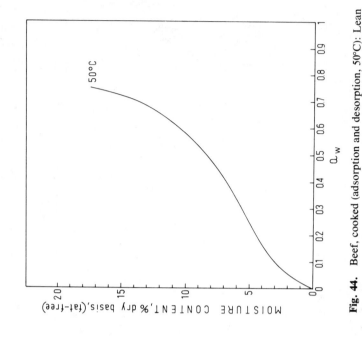

Fig. 44. Beef, cooked (adsorption and desorption, 50°C): Lean beef cooked in boiling water and cubes of 0.5 cm in size were air-dried at 55°C. Method: Static-desiccator (saturated salt/sulfuric acid solutions). (Iglesias and Chirife, 1976b).

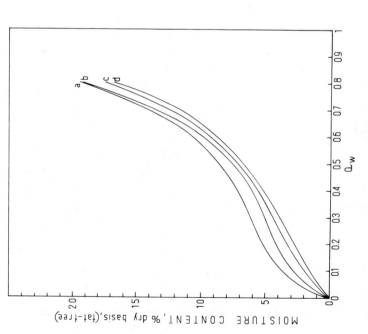

Fig. 43. Beef, cooked (adsorption, 30°C): Lean beef cooked in boiling water; cubes 0.5 cm in size were vacuum dried at 30°C and then heat-damaged in vacuum as follows: (b) 5 hr, 65°C; (c) 5 hr, 80°C; (d) 5 hr, 95°C. Method: Static-desiccator (saturated salt/sulfuric acid solutions). (Iglesias and Chirife, 1977.)

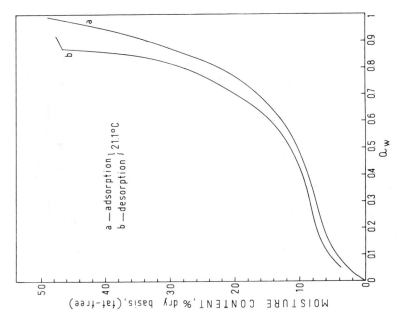

Fig. 46. Beef, precooked (adsorption and desorption, 21.1°C): Longissimus dorsi muscle of beef, ground, cooked, and freeze-dried at 40.6°C plate temperature. Method: Electrobalance assembly. (Palnitkar and Heldman, 1971.)

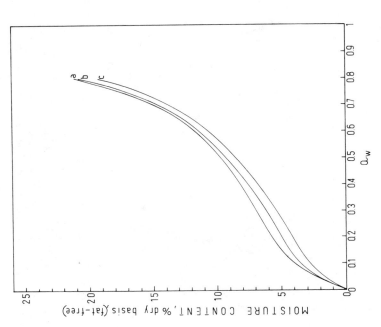

Fig. 45. Beef, cooked (adsorption, 30°C): Lean beef cooked in boiling water and cubes of 0.5 cm in size were air-dried as follows: (a) 30, (b) 55, (c) 70°C. Method: Static-desiccator (saturated salt/sulfuric acid solutions). (Iglesias and Chirife, 1976b).

33

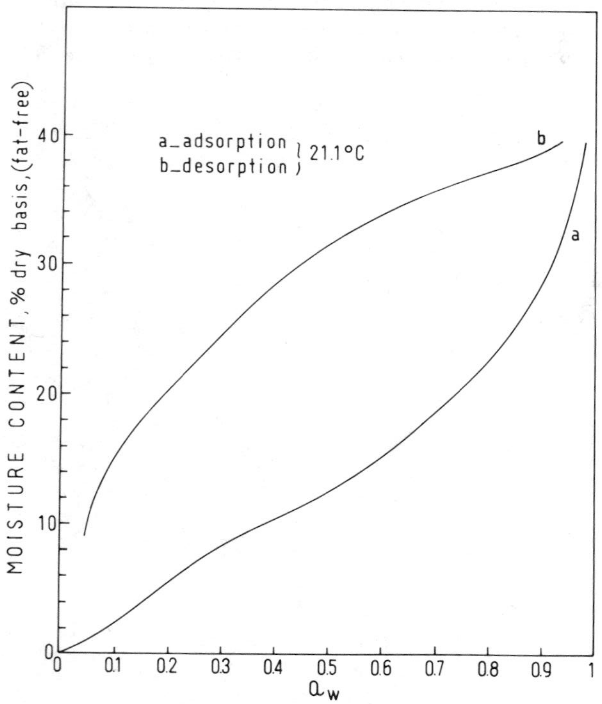

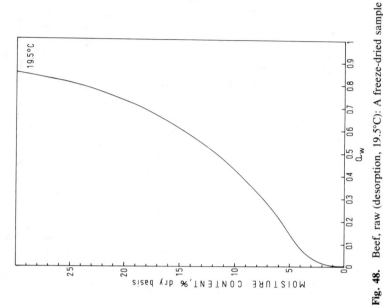

Fig. 48. Beef, raw (desorption, 19.5°C): A freeze-dried sample was ground and allowed to equilibrate to about 20% (dry basis) moisture content, before isotherm determination. Method: Manometric apparatus. (Taylor, 1961.)

Fig. 47. Beef, raw (adsorption and desorption, 21.1°C): Longissimus dorsi muscle of beef, ground and freeze-dried at 26.7°C. Method: Electrobalance assembly. (Palnitakar and Heldman, 1971.)

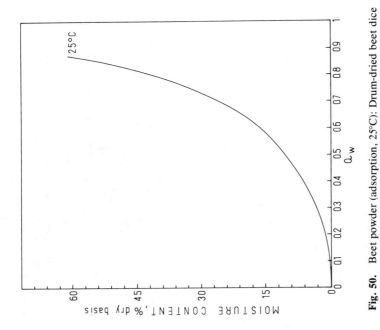

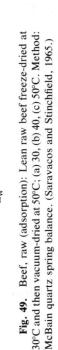

Fig. 49. Beef, raw (adsorption): Lean raw beef freeze-dried at 30°C and then vacuum-dried at 50°C: (a) 30, (b) 40, (c) 50°C. Method: McBain quartz spring balance. (Saravacos and Stinchfield, 1965.)

Fig. 50. Beet powder (adsorption, 25°C): Drum-dried beet dice (4% moisture) ground to pass through a U.S. #60 sieve. Method: Electrobalance assembly. (Kopelman and Saguy, 1977.)

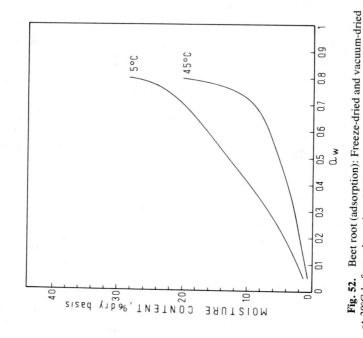

Fig. 52. Beet root (adsorption): Freeze-dried and vacuum-dried at 30°C before adsorption. Method: Jar with air agitation (sulfuric acid solutions). (Wolf *et al.*, 1973.)

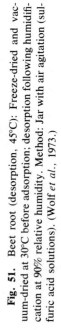

Fig. 51. Beet root (desorption, 45°C): Freeze-dried and vacuum-dried at 30°C before adsorption; desorption following humidification at 90% relative humidity. Method: Jar with air agitation (sulfuric acid solutions). (Wolf *et al.*, 1973.)

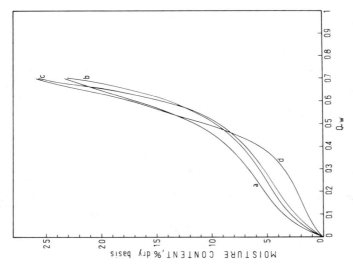

Fig. 54. Beet, sugar beet root (desorption): Mean sugar content of beets was 18.5% (wet basis); (a) 20, (b) 35, (c) 47, (d) 65°C. Method: Static-desiccator (saturated salt/sulfuric acid solutions). (Iglesias *et al.*, 1975a.)

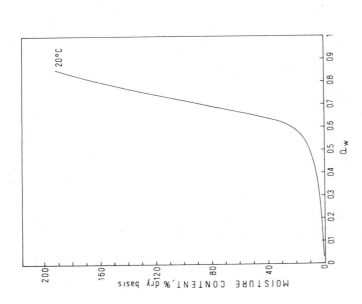

Fig. 53. Beet root juice (adsorption, -20°C): Spray dried. Method: Static-desiccator (saturated salt/sulfuric acid solutions). (Lewicki and Brzozowski, 1973.)

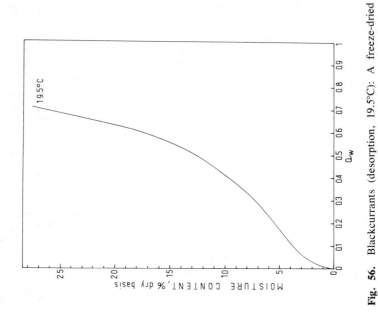

Fig. 56. Blackcurrants (desorption, 19.5°C): A freeze-dried sample was ground and allowed to equilibrate to about 20% (dry basis) moisture content before isotherm determination. Method: Manometric apparatus. (Taylor, 1961.)

Fig. 55. Beet, sugar beet root, water-insoluble components (adsorption and desorption): Samples 0.15 cm thick were leached with distilled water and freeze-dried at a plate temperature of 37°C. Method: Static-desiccator (saturated salt/sulfuric acid solutions). (Iglesias *et al.*, 1975a.)

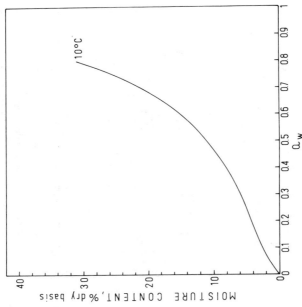

Fig. 58. Broccoli (adsorption, 10°C): Scalded and then freeze-dried; soluble constituents: 36.2% (dry basis). Method: Static-jar (sulfuric acid solutions). (Gane, 1950.)

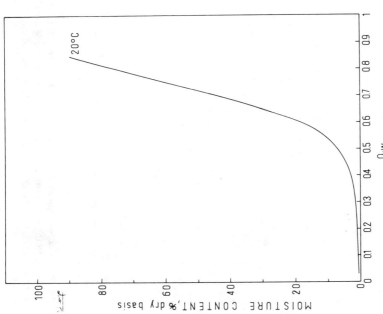

Fig. 57. Borsch, Instant (adsorption, 20°C). Method: Static-desiccator (saturated salt/sulfuric acid solutions). (Lewicki and Brzozowski, 1973.)

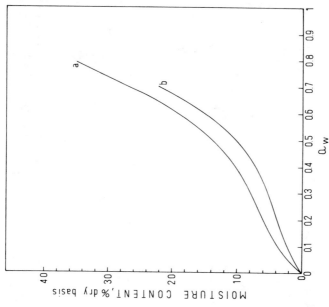

Fig. 60. Cabbage (adsorption and desorption): (a) A freeze-dried sample was ground and allowed to equilibrate to about 20% (dry basis) moisture content before isotherm determination; desorption, 19.5°C; (b) variety Savoy; samples for adsorption vacuum-dried at 70°C; adsorption, 37°C. Method: (a) manometric apparatus; (b) static-desiccator (sulfuric acid solutions). [(a) Taylor (1961); (b) Makower and Dehority (1943).]

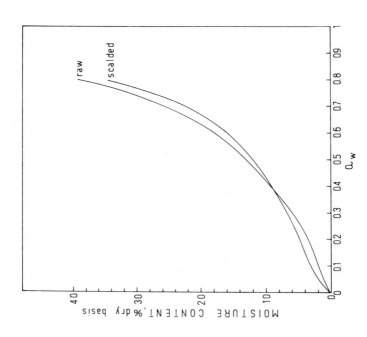

Fig. 59. Cabbage (adsorption, 10°C): Soluble constituents in raw material: 55.0% (dry basis); soluble constituents in scalded material: 42.1% (dry basis). Method: Static-jar (sulfuric acid solutions). (Gane, 1950.)

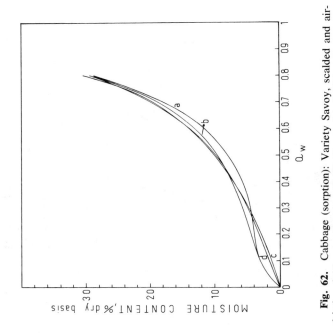

Fig. 62. Cabbage (sorption): Variety Savoy, scalded and air-dried, soluble constituents: 42.1% (dry basis); initial moisture content 4.7% (dry basis); (a) 0, (b) 10, (c) 25, (d) 37°C. Method: Static-jar (sulfuric acid solutions). (Gane, 1950.)

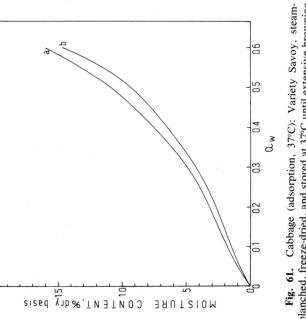

Fig. 61. Cabbage (adsorption, 37°C): Variety Savoy, steam-blanched, freeze-dried, and stored at 37°C until extensive browning occurred; (a) prior to, (b) after storage. Method: Static-desiccator (saturated salt solutions). (Mizrahi *et al.*, 1970.)

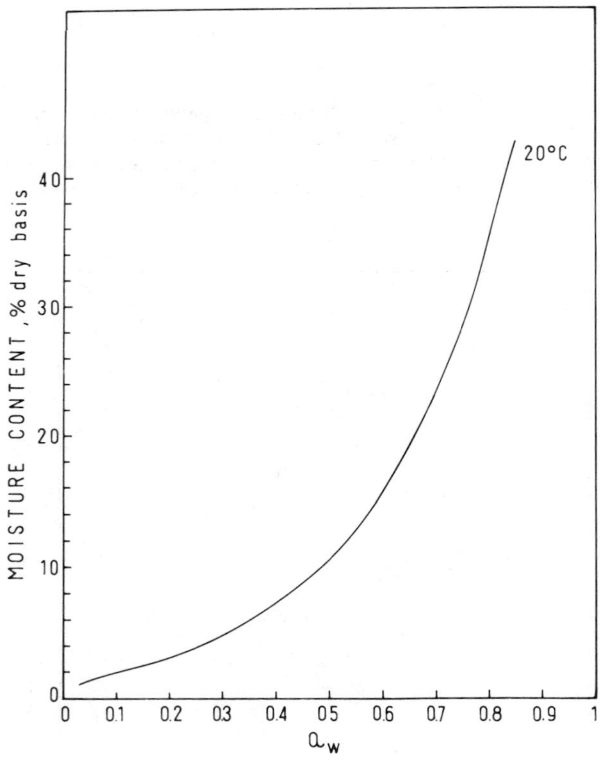

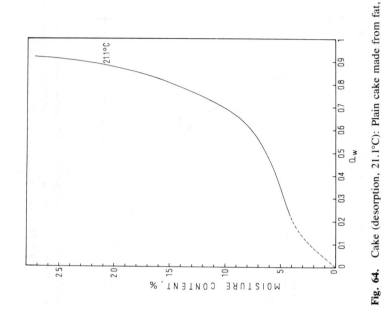

Fig. 63. Cabbage, dried, (adsorption, 20°C): Dried cabbage sample was ground for isotherm determination. Method: Static-desiccator (saturated salt/sulfuric acid solutions). (Lewicki and Brzozowski, 1973.)

Fig. 64. Cake (desorption, 21.1°C): Plain cake made from fat, sugar, egg, and flour; moisture content basis (wet or dry) not specified. Method: Landrock and Proctor (1951).

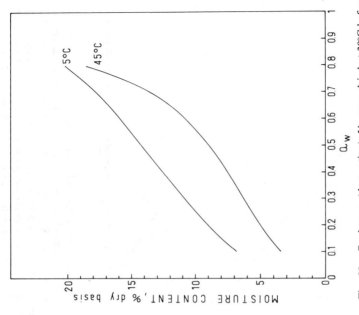

Fig. 66. Cardamom (desorption): Vacuum-dried at 30°C before adsorption; desorption following humidification at 90% relative humidity. Method: Jar with air agitation (sulfuric acid solutions). (Wolf *et al.*, 1973.)

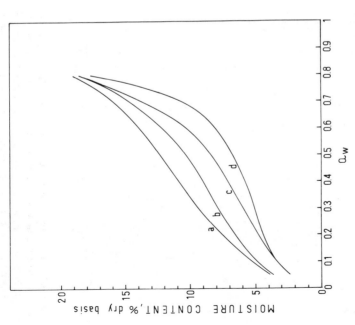

Fig. 65. Cardamom (adsorption): Vacuum-dried at 30°C before adsorption: (a) 5, (b) 25, (c) 45, (d) 60°C. Method: Jar with air agitation (sulfuric acid solutions). (Wolf *et al.*, 1973.)

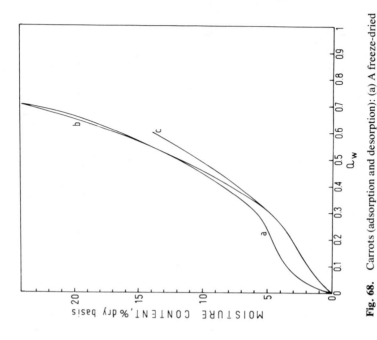

Fig. 68. Carrots (adsorption and desorption): (a) A freeze-dried sample was ground and allowed to equilibrate to about 20% (dry basis) moisture content before isotherm determination; desorption, 19.5°C; (b, c) variety Chantenay; samples for adsorption vacuum-dried at 70°C; adsorption and desorption, (b) 37, (c) 70°C. Method: (a) manometric apparatus; (b, c) static-desiccator (sulfuric acid solutions). [(a) Taylor (1961); (b, c) Makower and Dehority (1943).]

Fig. 67. β-Carotene–cellulose model (37°C): Microcrystalline cellulose dispersed in water and vacuum-dried at 50°C and β-carotene added; sodium metabisulfite used for sulfited samples; moisture content basis (wet or dry) not specified. Method: Static-desiccator (saturated salt solutions). (Baloch *et al.*, 1977.)

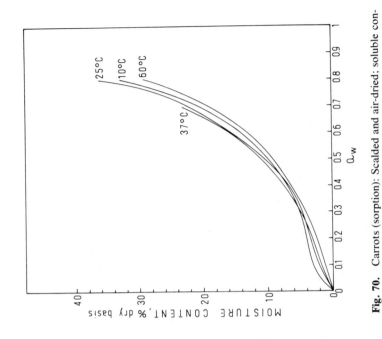

Fig. 70. Carrots (sorption): Scalded and air-dried; soluble constituents: 59.0% (dry basis); initial moisture content 9.4% (dry basis); (a) 37, (b) 25, (c) 10, (d) 60°C. Method: Static-jar (sulfuric acid solutions). (Gane, 1950.)

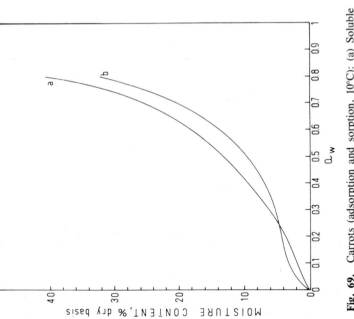

Fig. 69. Carrots (adsorption and sorption, 10°C): (a) Soluble constituents in raw (freeze-dried) material: 60.5% (dry basis): adsorption: (b) soluble constituents in scalded (spray-dried) material: 59.0% (dry basis); initial moisture content 5.7% (dry basis); sorption. Method: Static-jar (sulfuric acid solutions). (Gane, 1950.)

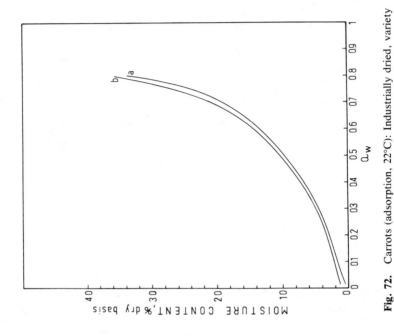

Fig. 72. Carrots (adsorption, 22°C): Industrially dried, variety Nantejska; (a) cabinet, (b) conveyor dryer. Method: Static-desiccator (saturated salt/sulfuric acid solutions). (Lewicki and Lenart, 1975.)

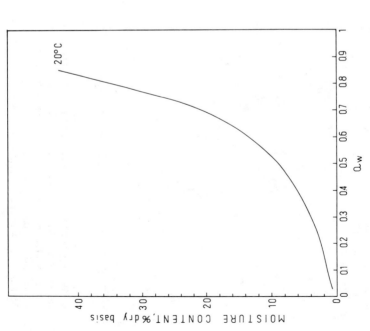

Fig. 71. Carrots, dried (adsorption, 20°C): The air-dried product was ground for isotherm determination. Method: Static-desiccator (saturated salt/sulfuric acid solutions). (Lewicki and Brzozowski, 1973.)

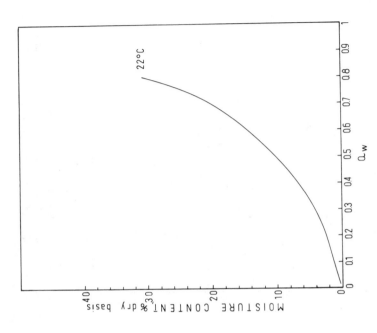

Fig. 73. Carrots (adsorption, 22°C): Laboratory-dried, blanched, variety Nantejska. Method: Static-desiccator (saturated salt/sulfuric acid solutions). (Lewicki and Lenart, 1975.)

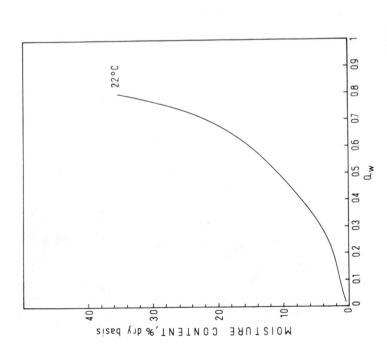

Fig. 74. Carrots (adsorption, 22°C): Laboratory-dried, cooked, variety Nantejska. Method: Static-desiccator (saturated salt/sulfuric acid solutions). (Lewicki and Lenart, 1975.)

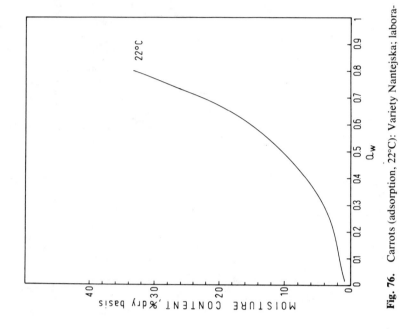

Fig. 76. Carrots (adsorption, 22°C): Variety Nantejska: laboratory-dried, sprayed with potato starch before drying. Method: Static-desiccator (saturated salt/sulfuric acid solutions). (Lewicki and Lenart, 1975.)

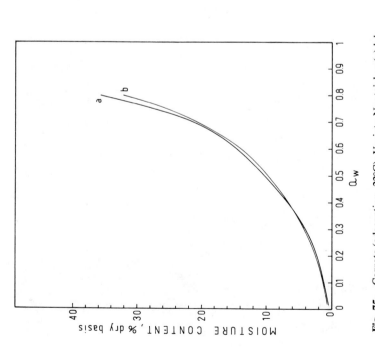

Fig. 75. Carrots (adsorption, 22°C): Variety Nantejska: (a) laboratory dried, frozen; (b) laboratory dried, dry–blanch–dry. Method: Static-desiccator (saturated salt/sulfuric acid solutions). (Lewicki and Lenart, 1975.)

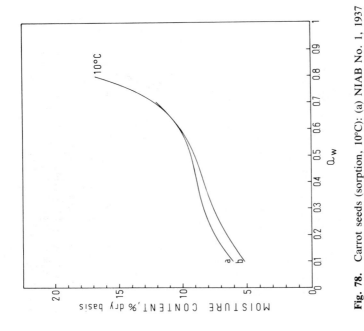

Fig. 78. Carrot seeds (sorption, 10°C): (a) NIAB No. 1, 1937 crop: initial moisture content 8.7% (dry basis); (b) Sutton's Chantenay, 1939 crop: initial moisture content 7.5% (dry basis). Method: Static-jar (sulfuric acid solutions). (Gane, 1948.)

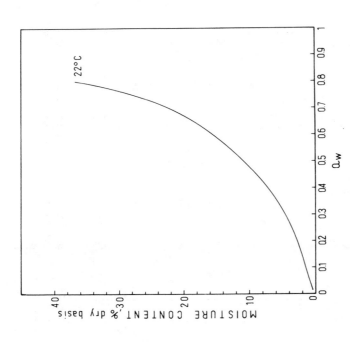

Fig. 77. Carrots (adsorption, 22°C): Variety Nantejska; laboratory-dried, frozen after the constant-rate drying period. Method: Static-desiccator (saturated salt/sulfuric acid solutions). (Lewicki and Lenart, 1975.)

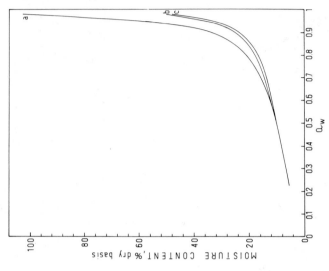

Fig. 80. Casein, Micellar (adsorption, 25°C): Micellar casein separated from milk of Simmental cows by centrifugation at 100,-000 g at 18°C for 2 hr; pH: (a) 4.7, (b) 7.8, (c) 6. Method: Isopiestic (saturated salt or sulfuric acid solutions). (Rüegg and Blanc, 1976.)

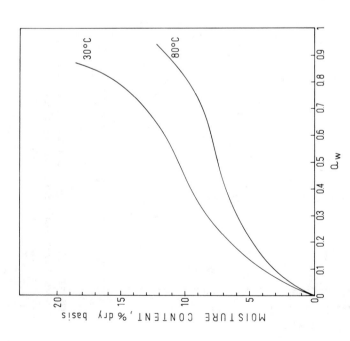

Fig. 79. Casein: Direction of sorption not specified. Method: Dynamic (air flow of known temperature and relative humidity). (Loncin *et al.*, 1968.)

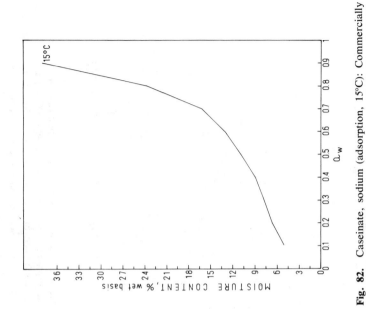

Fig. 81. Casein, Whole (adsorption, 25°C): Prepared by acid precipitation from milk of Simmental cows; pH (a) 7.75, (b) 6.1, (c) 4.6 (isoelectric point). Method: Isopiestic (saturated salt or sulfuric acid solutions). (Rüegg and Blanc, 1976.)

Fig. 82. Caseinate, sodium (adsorption, 15°C): Commercially available sodium caseinate (A/S Lidano): protein 91.5%, ash 4.6%, fat 1.1%. Method: Electrobalance assembly. (Hermansson, 1977.)

51

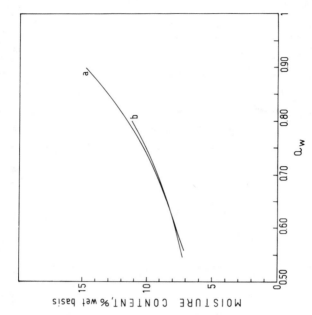

Fig. 83. Cashewnuts (sorption, 25°C): Cashewnuts from Brazil; (a) raw, (b) roasted, unsalted, (c) roasted, salted. Method: Static-desiccator (saturated salt/sulfuric acid solutions). (Quast and Teixeira Neto, 1976.)

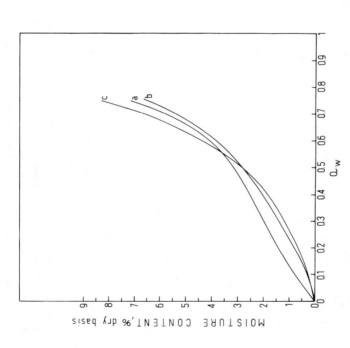

Fig. 84. Cashewnuts, whole (adsorption and desorption, 27°C): (a) Adsorption: nuts were air-dried at 40–45°C; (b) desorption: nuts were conditioned by exposure to the desired relative humidity. Method: Static-desiccator (saturated salt solutions). (Okwelogu and Mackay, 1969.)

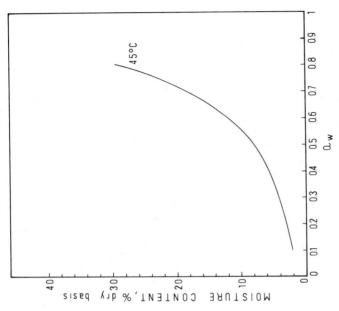

Fig. 86. Celery (desorption, 45°C): Freeze-dried and vacuum-dried at 30°C before desorption; desorption following humidification at 90% relative humidity. Method: Jar with air agitation (sulfuric acid solutions). (Wolf *et al.*, 1973.)

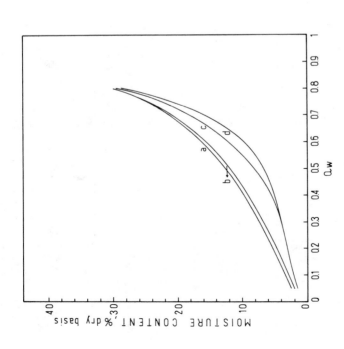

Fig. 85. Celery (adsorption): Freeze-dried and vacuum-dried at 30°C before adsorption; (a) 5, (b) 25, (c) 45, (d) 60°C. Method: Jar with air agitation (sulfuric acid solutions). (Wolf *et al.*, 1973.)

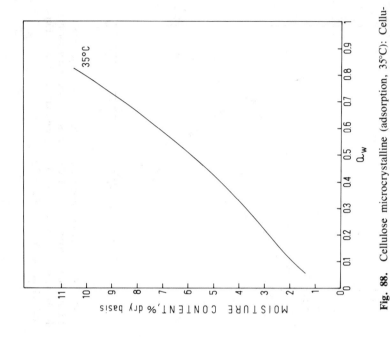

Fig. 88. Cellulose microcrystalline (adsorption, 35°C): Cellulose microcrystalline ("Avicel") dispersed in water and freeze-dried at room temperature before isotherm determination. Method: Static-desiccator (saturated salt solutions). (Iglesias *et al.*, 1980.)

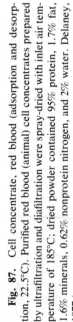

Fig. 87. Cell concentrate, red blood (adsorption and desorption, 22.5°C). Purified red blood (animal) cell concentrates prepared by ultrafiltration and diafiltration were spray-dried with inlet air temperature of 185°C; dried powder contained 95% protein, 1.7% fat, 1.6% minerals, 0.62% nonprotein nitrogen, and 2% water. Delaney, 1977.)

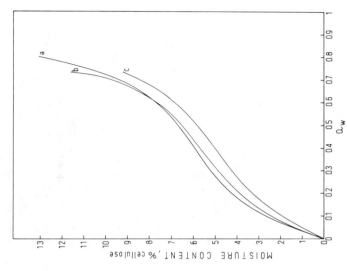

Fig. 90. Cellulose microcrystalline-oil model (desorption, 37°C): Cellulose-based model (with added surfactant "Span 20") system of the following composition: "Avicel" (cellobiose particles 1 μm) ~ 93% (dry basis), "Apiezon B" oil ~ 7% (dry basis); (a) control, (b) with 0.75% "Span 20," (c) with 7.5% "Span 20." Method: Static-desiccator (saturated salt solutions). (Labuza and Rutman, 1968.)

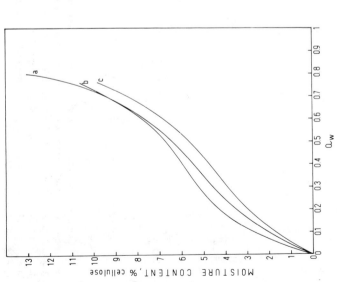

Fig. 89. Cellulose microcrystalline-oil model (desorption, 37°C): Cellulose-based model (with added surfactant "Tween 20") system of the following composition: "Avicel" (cellobiose particles 1 μm) ~ 93% (dry basis), "Apiezon B" oil ~ 7% (dry basis); (a) control, (b) with 0.75% "Tween 20," (c) with 7.5% "Tween 20." Method: Static-desiccator (saturated salt solutions). (Labuza and Rutman, 1968.)

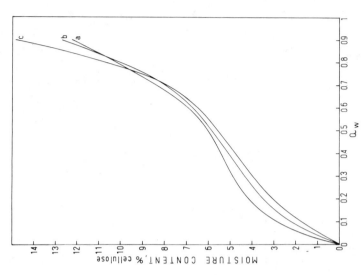

Fig. 92. Cellulose microcrystalline-oil model (adsorption, 37°C): Cellulose-based model system (with added surfactant "Tween 20") of the following composition: "Avicel" (cellobiose particles 1 μm) ∼93% (dry basis), "Apiezon B" oil ∼7% (dry basis); (a) control, (b) with 0.75% "Tween 20," (c) with 7.5% "Tween 20." Method: Static-desiccator (saturated salt solutions) (Labuza and Rutman, 1968.)

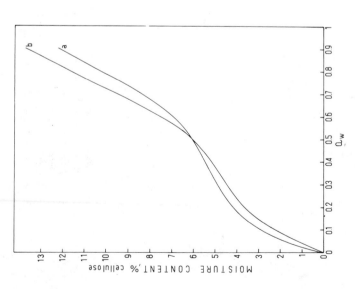

Fig. 91. Cellulose microcrystalline-oil model (adsorption, 37°C): Cellulose-based model system with added surfactant of the following composition: "Avicel" (cellobiose particles 1 μm) ∼93% (dry basis), "Apiezon B" oil ∼7% (dry basis); (a) control, (b) with 7.5% "Span 20." Method: Static-desiccator (saturated salt solutions). (Labuza and Rutman, 1968.)

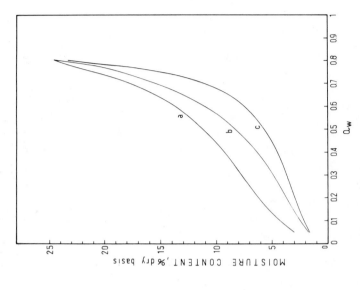

Fig. 94. Camomile tea (adsorption): Vacuum-dried at 30°C before adsorption: (a) 5 and 25, (b) 45, (c) 60°C. Method: Jar with air agitation (sulfuric acid solutions). (Wolf *et al.*, 1973.)

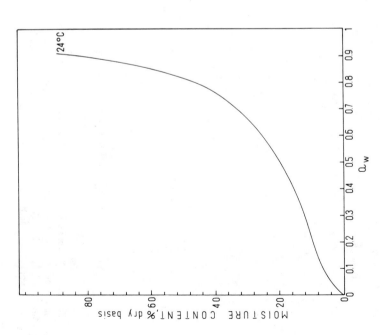

Fig. 93. Cellulose sodium carboxymethyl (adsorption, 24°C: "Hercules" sodium carboxymethyl cellulose CMC-7LF, spray-dried. Electrobalance assembly. (Berlin *et al.*, 1973b.)

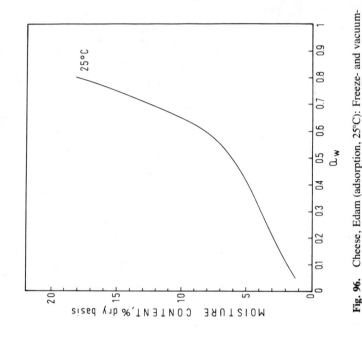

Fig. 96. Cheese, Edam (adsorption, 25°C): Freeze- and vacuum-dried at 30°C before adsorption. Method: Jar with air agitation (sulfuric acid solutions). (Wolf *et al.*, 1973.)

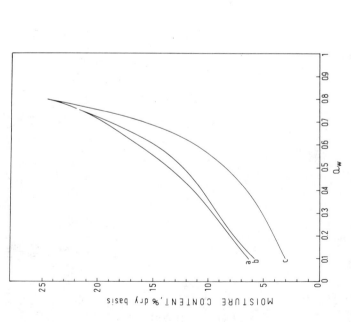

Fig. 95. Camomile tea (desorption): Vacuum-dried at 30°C before adsorption; desorption following humidification at 90% relative humidity; (a) 5, (b) 25, (c) 45°C. Method: Jar with air agitation (sulfuric acid solutions). (Wolf *et al.*, 1973.)

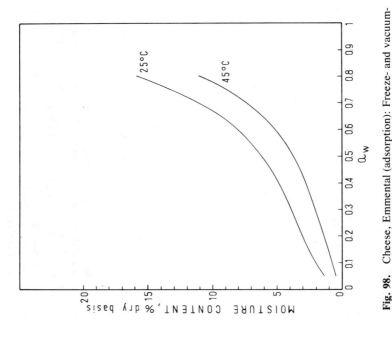

Fig. 98. Cheese, Emmental (adsorption): Freeze- and vacuum-dried at 30°C before adsorption. Method: Jar with air agitation (sulfuric acid solutions). (Wolf *et al.*, 1973.)

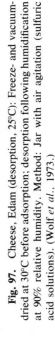

Fig. 97. Cheese, Edam (desorption, 25°C): Freeze- and vacuum-dried at 30°C before adsorption; desorption following humidification at 90% relative humidity. Method: Jar with air agitation (sulfuric acid solutions). (Wolf *et al.*, 1973.)

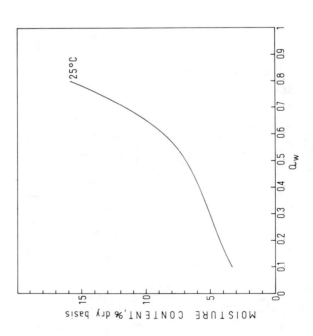

Fig. 100. Cheese and paracasein (adsorption and desorption, 20°C): Salted Gouda cheese contained 12.6 g NaCl/100 g protein and unsalted 0.4 g NaCl/100 g protein; paracasein was prepared from fresh skim milk by acidification at pH 4.6–4.8 (30–35°C); (a) paracasein, adsorption, (b) cheese, unsalted, desorption, (c) cheese, salted, desorption. Method: Static-desiccator (sulfuric acid/saturated salt solutions). (Geurts *et al.*, 1974.)

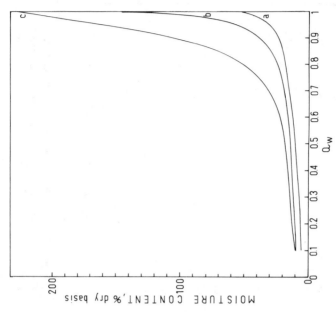

Fig. 99. Cheese, Emmental (desorption, 25°C): Freeze- and vacuum-dried at 30°C before adsorption; desorption following humidification at 90% relative humidity. Method: Jar with air agitation (sulfuric acid solutions). (Wolf *et al.*, 1973.)

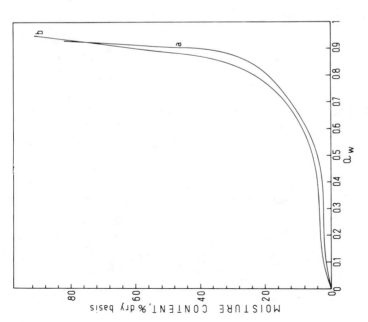

Fig. 102. Cheese, cottage, whey (adsorption and desorption, 24°C): Laboratory freeze-dried cottage cheese whey; desorption measured following equilibration at near saturation pressure; little crystalline α-lactose. Method: Electrobalance assembly. (Berlin and Anderson, 1975.)

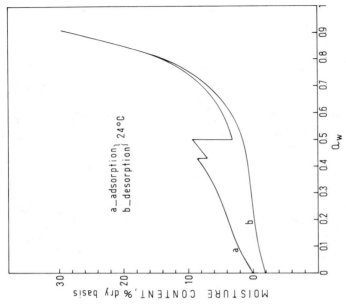

Fig. 101. Cheese, cottage, whey (desorption, 24°C): Desorption isotherm was obtained after equilibration at near saturation pressure; (a) 56.3% crystalline lactose, (b) 8.8% crystalline lactose. Method: Electrobalance assembly. (Berlin and Anderson, 1975.)

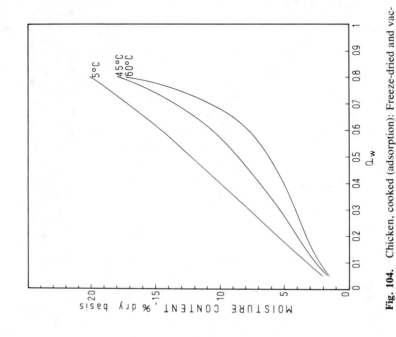

Fig. 104. Chicken, cooked (adsorption): Freeze-dried and vacuum-dried at 30°C before adsorption; (a) 5, (b) 45, (c) 60°C. Method: Jar with air agitation (sulfuric acid solutions). (Wolf *et al.*, 1973.)

Fig. 103. Chicken, cooked (desorption, 19.5°C): A freeze-dried sample was ground and allowed to equilibrate to a moisture content of about 20% (dry basis). Method: Manometric apparatus. (Taylor, 1961.)

Fig. 106. Chicken, raw (adsorption): Freeze-dried and vacuum-dried at 30°C before adsorption; (a) 5, (b) 45, (c) 60°C. Method: Jar with air agitation (sulfuric acid solutions). (Wolf *et al.*, 1973.)

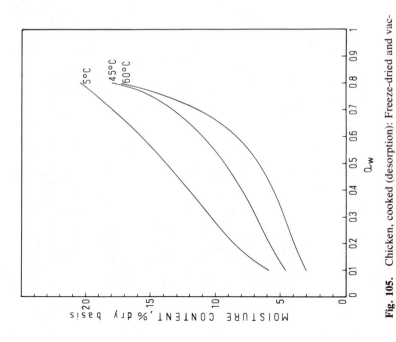

Fig. 105. Chicken, cooked (desorption): Freeze-dried and vacuum-dried at 30°C before adsorption; desorption following humidification at 90% relative humidity; (a) 5, (b) 45, (c) 60°C. Method: Jar with air agitation (sulfuric acid solutions). (Wolf *et al.*, 1973.)

63

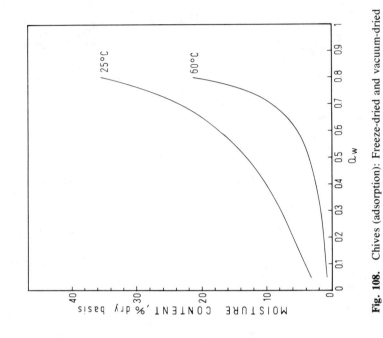

Fig. 108. Chives (adsorption): Freeze-dried and vacuum-dried at 30°C before adsorption. Method: Jar with air agitation (sulfuric acid solutions). (Wolf *et al.*, 1973.)

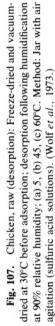

Fig. 107. Chicken, raw (desorption): Freeze-dried and vacuum-dried at 30°C before adsorption; desorption following humidification at 90% relative humidity; (a) 5, (b) 45, (c) 60°C. Method: Jar with air agitation (sulfuric acid solutions). (Wolf *et al.*, 1973.)

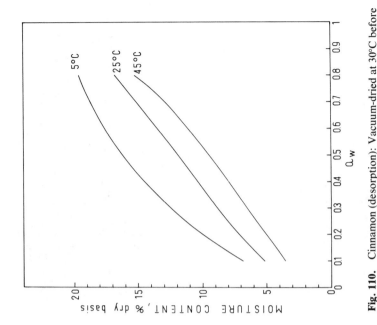

Fig. 110. Cinnamon (desorption): Vacuum-dried at 30°C before adsorption; desorption following humidification at 90% relative humidity; (a) 5, (b) 25, (c) 45°C. Method: Jar with air agitation (sulfuric acid solutions). (Wolf *et al.*, 1973.)

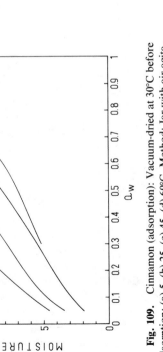

Fig. 109. Cinnamon (adsorption): Vacuum-dried at 30°C before adsorption; (a) 5, (b) 25, (c) 45, (d) 60°C. Method: Jar with air agitation (sulfuric acid solutions). (Wolf *et al.*, 1973.)

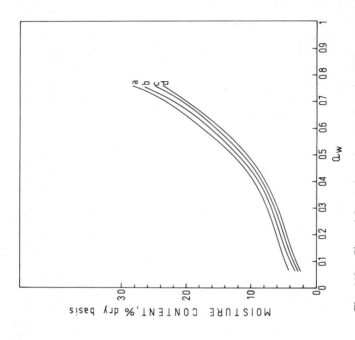

Fig. 112. Citrus juice (adsorption, 25°C). Spray-dried Citrus Unshiu juice with various contents of pulp: (a) 10, (b) 15, (c) 20, (d) 25%. Method: Static-desiccator (saturated salt solutions). (Komeyasu and Iyama, 1974.)

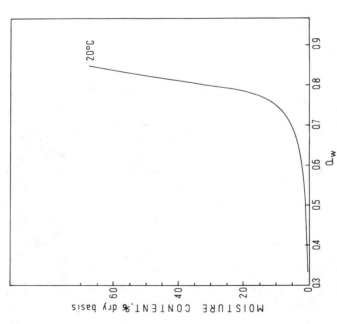

Fig. 111. Citric Acid (desorption, 20°C). Method: Static-desiccator (saturated salt/sulfuric acid solutions). (Lewicki and Brzozowski, 1973.)

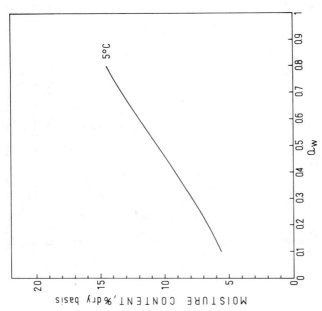

Fig. 114. Cloves (desorption, 5°C): Vacuum-dried at 30°C before adsorption; desorption following humidification at 90% relative humidity. Method: Jar with air agitation (sulfuric acid solutions). (Wolf *et al.*, 1973.)

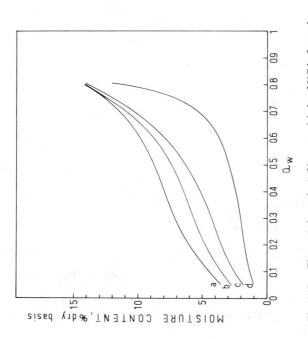

Fig. 113. Cloves (adsorption): Vacuum-dried at 30°C before adsorption; (a) 5, (b) 25, (c) 45, (d) 60°C. Method: Jar with air agitation (sulfuric acid solutions). (Wolf *et al.*, 1973.)

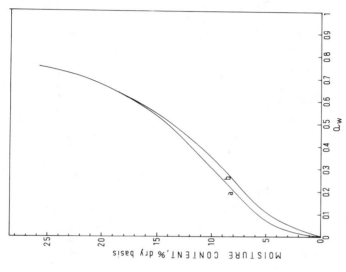

Fig. 116. Cod, raw [(a) desorption, 19.5°C; (b) adsorption, 30°C)]: (a) A freeze-dried sample was ground and allowed to equilibrate to about 20% (dry basis). Method: (a) Manometric apparatus; (b) static-jar (sulfuric acid solutions). [(a) Taylor (1961); (b) Jason (1958).]

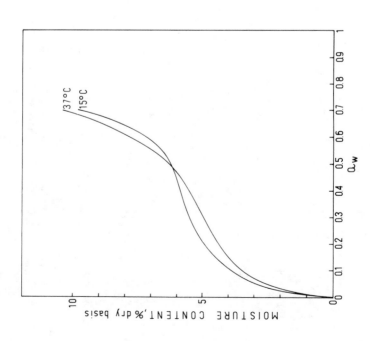

Fig. 115. Cocoa (sorption). Method: Static-jar (sulfuric acid solutions). (Gane, 1950.)

68

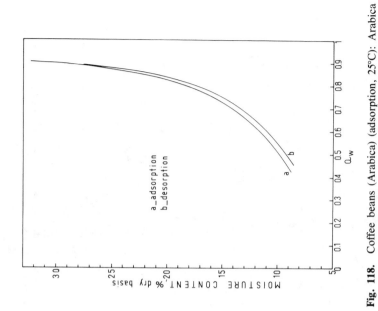

Fig. 118. Coffee beans (Arabica) (adsorption, 25°C): Arabica beans from Kenya, air-dried at room temperature prior to adsorption. Method: Dew-point. (Ayerst, 1965.)

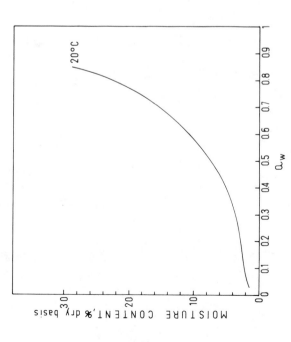

Fig. 117. Coffee (Inka) (adsorption, 20°C): Product based on grain and malt. Method: Static-desiccator (sulfuric acid/saturated salt solutions). (Lewicki and Brzozowski, 1973.)

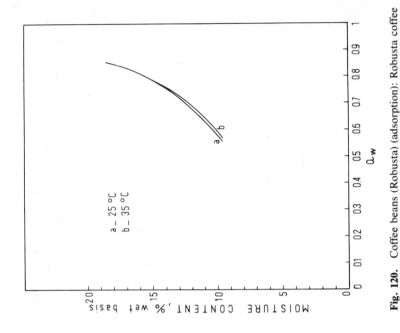

Fig. 120. Coffee beans (Robusta) (adsorption): Robusta coffee beans from Uganda, air-dried at room temperature before adsorption. Method: Dew-point. (Ayerst, 1965.)

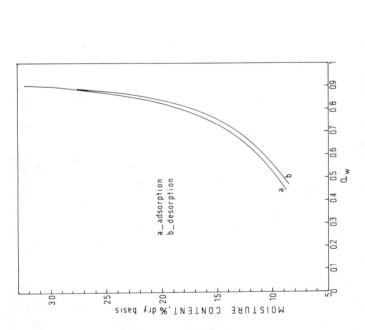

Fig. 119. Coffee beans (Arabica) (adsorption, 35°C): Arabica beans from Kenya, air-dried at room temperature prior to adsorption. Method: Dew-point. (Ayerst, 1965.)

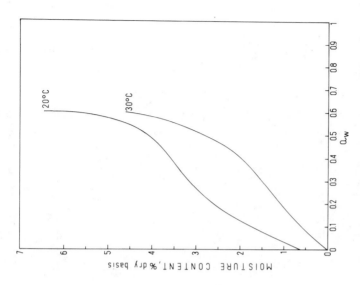

Fig. 122. Coffee, roasted and ground, decaffeinated (sorption). Method: Static-desiccator (saturated salt solutions). (Hayakawa *et al.*, 1978.)

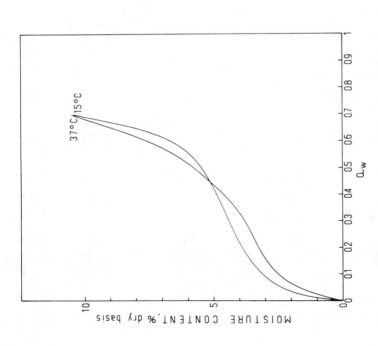

Fig. 121. Coffee, roasted and ground (sorption): Coffee was 60% Cameron, 40% Kenya. Method: Static-jar (sulfuric acid solutions). (Gane, 1950.)

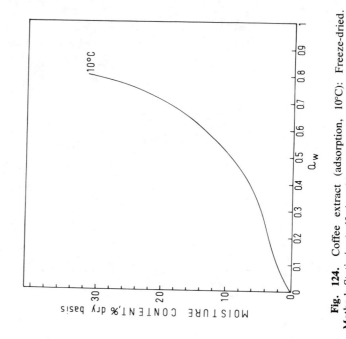

Fig. 124. Coffee extract (adsorption, 10°C): Freeze-dried. Method: Static-jar (sulfuric acid solutions). (Gane, 1950.)

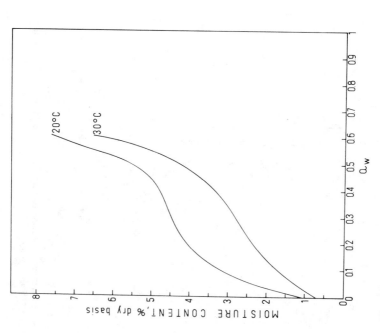

Fig. 123. Coffee, roasted and ground, undecaffeinated (sorption). Method: Static-desiccator (saturated salt solutions). (Hayakawa et al., 1978.)

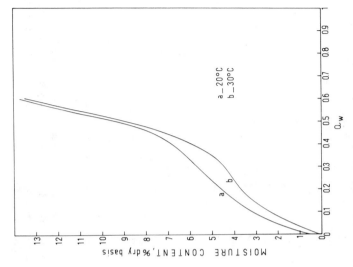

Fig. 125. Coffee extract, decaffeinated (adsorption): Freeze-dried; freshly prepared sample in a commercial processing plant. Method: Static-desiccator (saturated salt solutions). (Hayakawa *et al.*, 1978.)

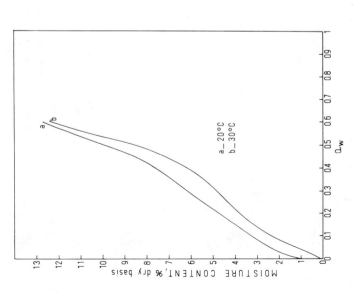

Fig. 126. Coffee extract, undecaffeinated (adsorption): Freeze-dried; freshly prepared sample in a commercial processing plant. Method: Static-desiccator (saturated salt solutions). (Hayakawa *et al.*, 1978.)

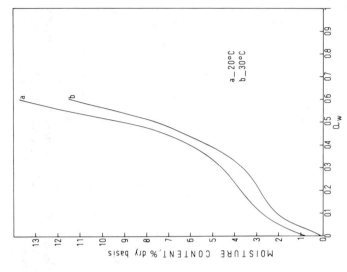

Fig. 128. Coffee extract, agglomerated, undecaffeinated (sorption): Spray-dried; freshly prepared sample in a commercial processing plant. Method: Static-desiccator (saturated salt solutions). (Hayakawa *et al.*, 1978.)

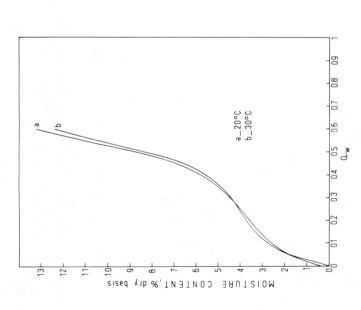

Fig. 127. Coffee extract, agglomerated, decaffeinated (sorption): Spray-dried; freshly prepared sample in a commercial processing plant. Method: Static-desiccator (saturated salt solutions). (Hayakawa *et al.*, 1978.)

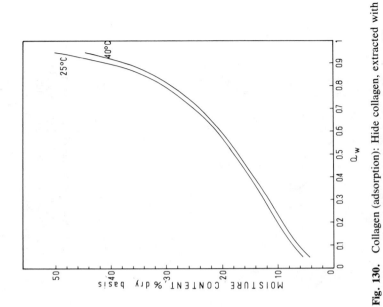

Fig. 129. Coffee products (sorption, 28°C): Brazilian coffee products; (a) soluble coffee, (b) green coffee, (c) roasted coffee, 11% weight loss, (d) roasted coffee, 19% weight loss, (e) spent coffee grounds. Method: Static-desiccator (saturated salt/sulfuric acid solutions). (Quast and Teixeira Neto, 1976.)

Fig. 130. Collagen (adsorption): Hide collagen, extracted with ether and alcohol, washed (dilute HCl), electrodialyzed, and freeze-dried. Method: Static-desiccator (sulfuric acid solutions). (Bull, 1944.)

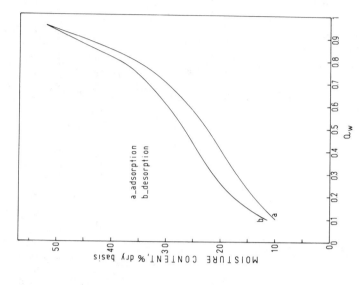

Fig. 132. Collagen, chrome-tanned: Temperature of adsorption and desorption not specified; collagen fibers teased out from cow hide belly portion; washed in 50% acetone and in acetone and tanned with chrome. (Sanjeevi and Ramanathan, 1976.)

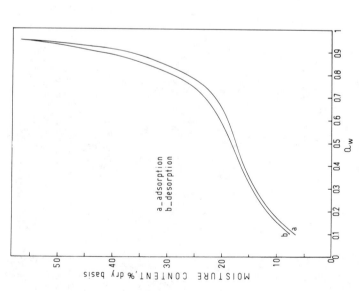

Fig. 131. Collagen: Temperature of adsorption and desorption not specified; collagen fibers teased out from cow hide belly portion; washed in 50% acetone and then in acetone. (Sanjeevi and Ramanathan, 1976.)

76

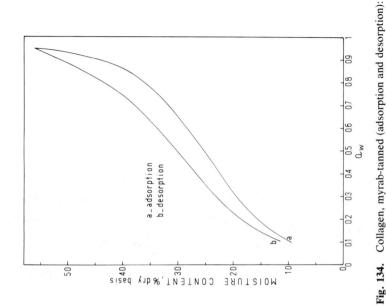

Fig. 133. Collagen, enzyme treated (adsorption and desorption): Temperature of adsorption and desorption not specified; collagen fibers teased out from cow hide belly portion, washed in 50% acetone and then in acetone, and subjected to enzyme treatment to remove noncollagenous components. (Sanjeevi and Ramanathan, 1976.)

Fig. 134. Collagen, myrab-tanned (adsorption and desorption): Temperature of adsorption and desorption not specified; collagen fibers teased out from cow hide belly portion, washed in 50% acetone and in acetone, and myrab tanned. (Sanjeevi and Ramanathan, 1976.)

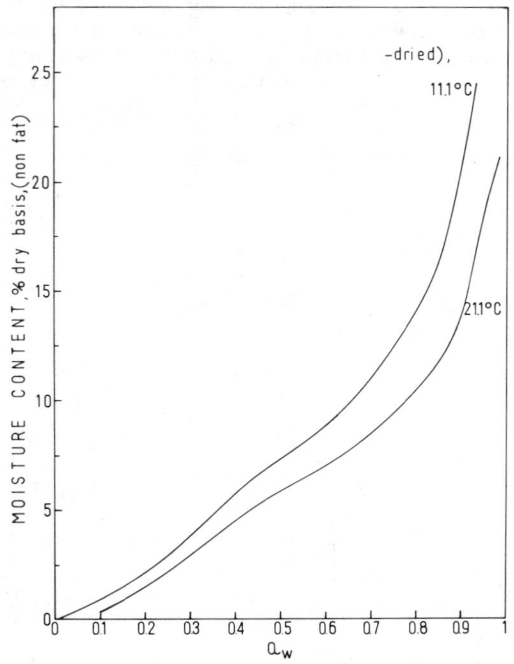

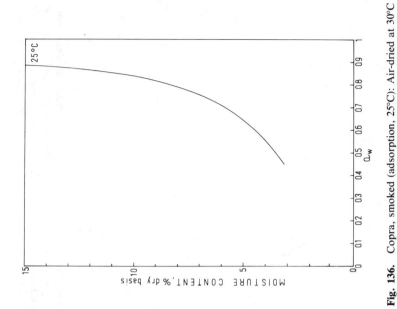

Fig. 135. Connective tissue, beef (adsorption): Extracted from longissimus dorsi muscle; final white fibrous mass freeze-dried at 26.7°C plate temperature. Method: Electrobalance assembly. (Palnitkar and Heldman, 1971.)

Fig. 136. Copra, smoked (adsorption, 25°C): Air-dried at 30°C to an $a_w = 0.15$–0.20 before adsorption. Method: Dew-point. (Pixton and Warburton, 1971b.).

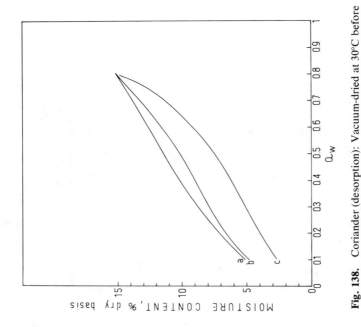

Fig. 138. Coriander (desorption): Vacuum-dried at 30°C before adsorption; desorption following humidification at 90% relative humidity; (a) 5, (b) 25, (c) 45°C. Method: Jar with air agitation (sulfuric acid solutions). (Wolf *et al.*, 1973.)

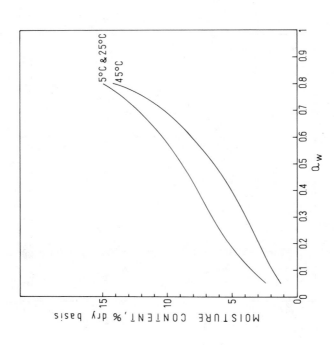

Fig. 137. Coriander (adsorption): Vacuum-dried at 30°C before adsorption; (a) 5, 25, (b) 45°C. Method: Jar with air agitation (sulfuric acid solutions). (Wolf *et al.*, 1973.)

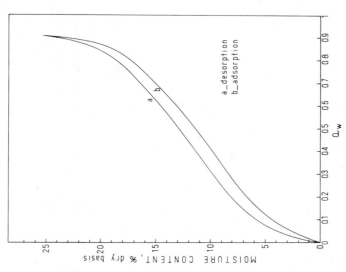

Fig. 140. Corn (adsorption and desorption, 30°C): Values are averages for varieties Schwenk 13 (1953 crop, sound, whole kernels), Schwenk 13 (1953 crop ground, 20-mesh), Dyar 444, and Illinois 1277; for adsorption, samples were dried in a vacuum oven at 72–76°C; for desorption samples were first exposed to 97% relative humidity. Method: Static-desiccator (saturated salt solutions). (Hubbard *et al.*, 1957.)

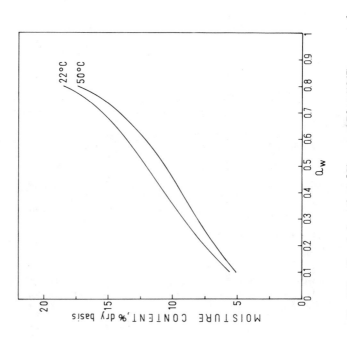

Fig. 139. Corn (adsorption). [Chung and Pfost (1967), as quoted by Ngoddy and Bakker-Arkema (1970).]

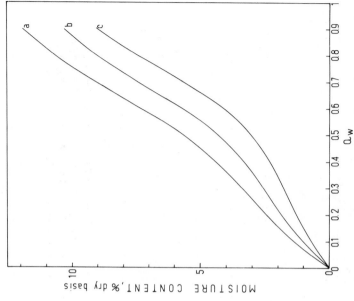

Fig. 142. Corn flours (adsorption, 25°C): Corn was a Whisnand Hybrid (Whisnand Seed Co., Arcola, IL); soybean was of the Wayne variety; gelatinization was done at 82°C and drying on a double-drum drier; (a) pregelled corn flour, (b) 50% pregelled corn flour, 50% full fat soy flour, (c) full fat soy flour. Method: Static-desiccator (saturated salt solutions). (Ayernor and Steinberg, 1977.)

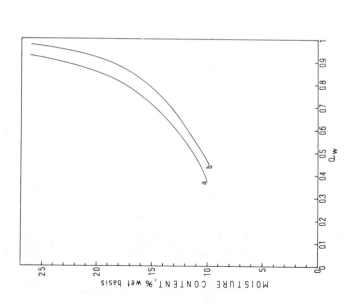

Fig. 141. Corn (adsorption, 25°C): (a) English, (b) American. Method: Dew-point. (Pixton and Warburton, 1971a.)

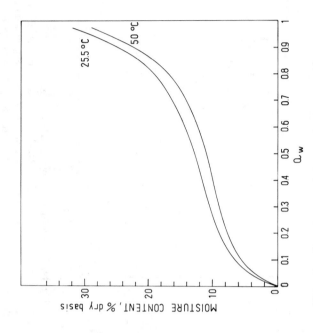

Fig. 143. Corn flour, degermed (adsorption): The pedigree of Midwest corn variety was (WF9 MST × H71 × 0.1 43RFXB 37RF); samples were air-dried at 75°C prior to adsorption. Method: Static-desiccator (saturated salt solutions). (Kumar, 1974.)

Fig. 144. Corn flour, whole (adsorption): The pedigree of Midwest corn variety was (WF9 MST × H71 × 0.1 43RFXB 37RF); samples were air-dried at 75°C prior to adsorption. Method: Static-desiccator (saturated salt solutions). (Kumar, 1974.)

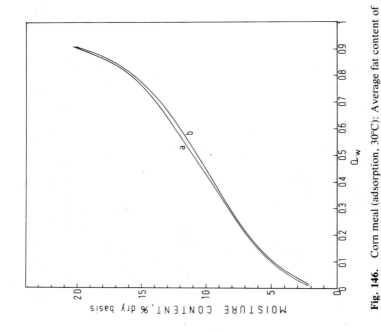

Fig. 146. Corn meal (adsorption, 30°C): Average fat content of sifted (b) and special-sifted (a) maize meals was 3.7 and 2.6%, respectively; average fiber content of sifted and special-sifted maize meals was 0.96 and 0.74%, respectively. Method: Jar with air agitation (sulfuric acid/saturated salt solutions). (van Twisk, 1969.)

Fig. 145. Corn germ flour (adsorption): The pedigree of Midwest corn variety was (WF9 MST × H71 × 0.1 43RFXB 37RF); samples were air-dried at 75°C prior to adsorption. Method: Static-desiccator (saturated salt solutions). (Kumar, 1974.)

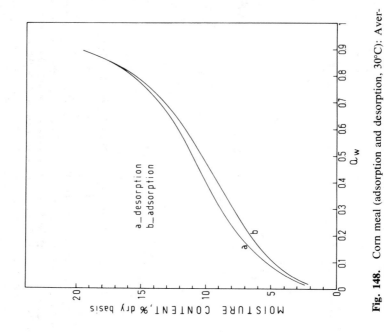

Fig. 148. Corn meal (adsorption and desorption, 30°C): Average fat and fiber content of unsifted maize meal was 5 and 1.9%, respectively. Method: Jar with air agitation (sulfuric acid/saturated salt solutions). (van Twish, 1969.)

Fig. 147. Corn meal (desorption, 30°C): Average fat content of sifted (b) and special-sifted (a) maize meals was 3.7 and 2.6%, respectively; average fiber content of sifted and special-sifted maize meals was 0.96 and 0.74%, respectively. Method: Jar with air agitation (sulfuric acid/saturated salt solutions). (van Twisk, 1969.)

84

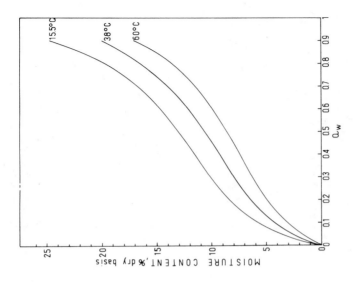

Fig. 150. Corn, shelled (desorption): (a) 15.5, (b) 38, (c) 60°C. [Rodriguez Arias (1956), as quoted by Chen and Clayton (1971).]

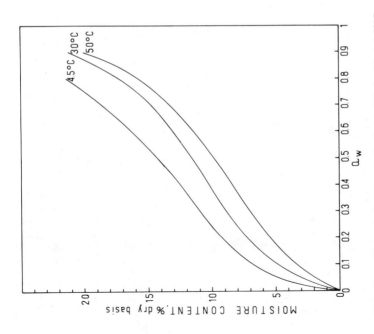

Fig. 149. Corn, shelled (desorption): (a) 4.5, (b) 30, (c) 50°C. [Rodriguez Arias (1956), as quoted by Chen and Clayton (1971).]

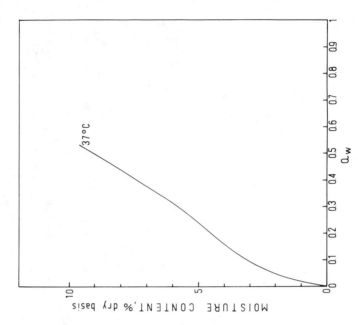

Fig. 152. Cottonseed meal (sorption, 37°C): Extracted, deglanded cottonseed flour (PRO-FAM C-650, Grain Processing Corp., IO). Method: Static-desiccator (saturated salt solutions). (Mizrahi and Karel, 1977.)

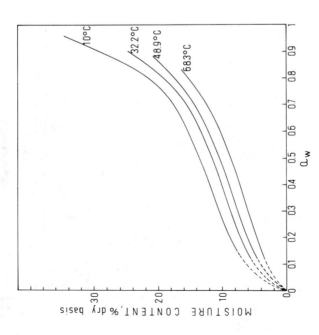

Fig. 151. Corn, shelled (desorption): Grain used was De Kalb XL-66 (crop 1971); (a) 10, (b) 32.2, (c) 48.9, (d) 68.3°C. Method: Desiccator with air agitation (saturated salt solutions). (Gustafson and Hall, 1974.)

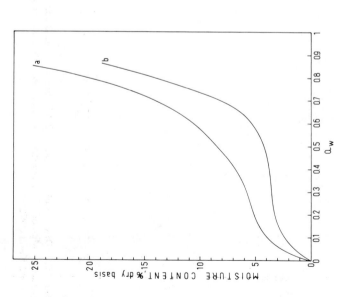

Fig. 154. Curd gels (adsorption and desorption, 20°C): Normal curd (dahi) gel was prepared from reconstituted whole spray-dried milk heated at 85°C, cooled, and inoculated with *S. thermophilus*; curd powder was prepared by freeze-drying fresh curd. Normal curd gel: (a) adsorption, (b) desorption; curd powder; (c) adsorption, (d) desorption. Method: Isopiestic. (Baisya *et al.*, 1975.)

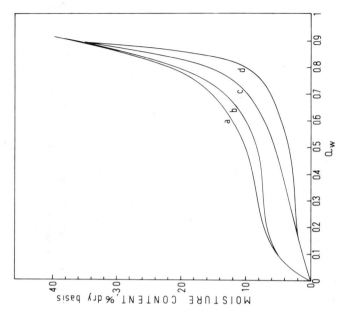

Fig. 153. Curd gels (adsorption, 50°C): Normal curd (dahi) gel (a) was prepared from reconstituted spray-dried whole milk heated at 85°C, cooled, and inoculated with *S. thermophilus*; curd powder (b) was prepared by freeze-drying fresh curd. Method: Isopiestic. (Baisya *et al.*, 1975.)

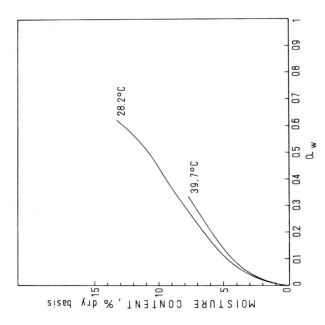

Fig. 156. Dextrin (adsorption): Commercial white dextrin from corn starch, Allied Chemical and Dye Corp., lot 14. (Volman *et al.*, 1960.)

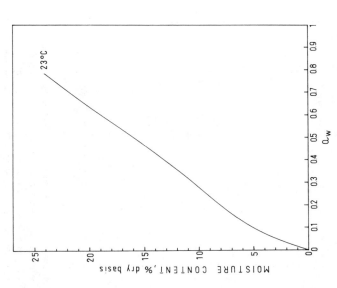

Fig. 155. Dextran-10 (adsorption, 23°C): Freeze-dried at room temperature. Method: Static-desiccator (saturated salt solutions). (Flink and Karel, 1972.)

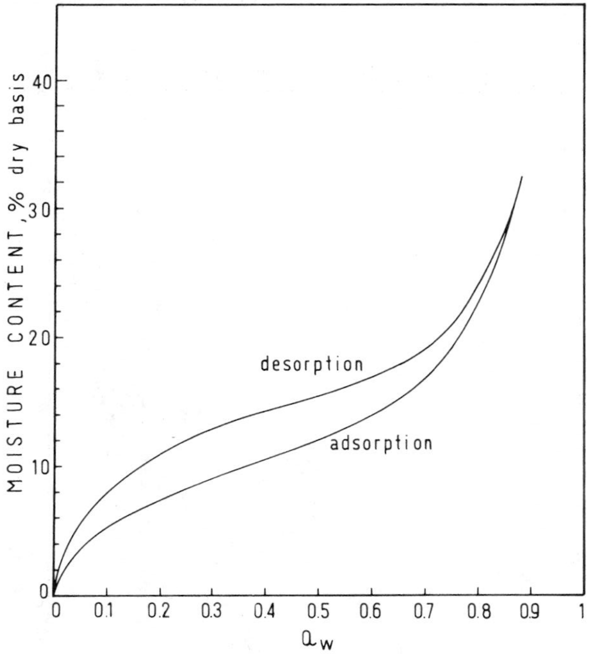

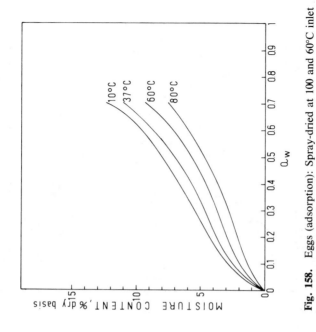

Fig. 157. Dextrin (adsorption and desorption, 10.7°C): Commercial white dextrin from corn starch, Allied Chemical and Dye Corp., lot 14. (Volman et al., 1960.)

Fig. 158. Eggs (adsorption): Spray-dried at 100 and 60°C inlet and outlet air temperature, respectively; (a) 10, (b) 37, (c) 60, (d) 80°C. Method: Static-jar (sulfuric acid solutions). (Gane, 1943.)

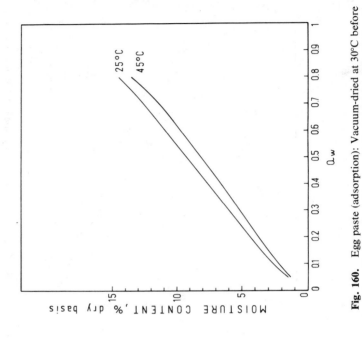

Fig. 160. Egg paste (adsorption): Vacuum-dried at 30°C before adsorption. Method: Jar with air agitation (sulfuric acid solutions). (Wolf *et al.*, 1973.)

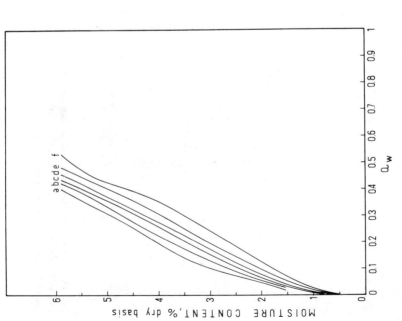

Fig. 159. Eggs, dehydrated (adsorption): Commercial spray-dried eggs; (a) 17.1, (b) 30, (c) 40, (d) 50, (e) 60, (f) 70°C. Method: Static-desiccator (sulfuric acid solutions). (Makower, 1943.)

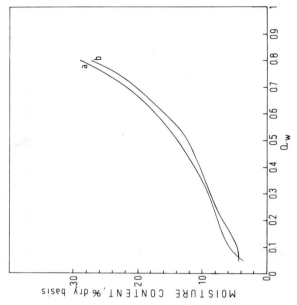

Fig. 162. Egg white and yolk (adsorption, 10°C): Spray-dried with 100°C inlet and 60°C outlet air temperature: (a) egg white, (b) egg yolk, fat-free. **Method:** Static-jar (sulfuric acid solutions). (Gane, 1943.)

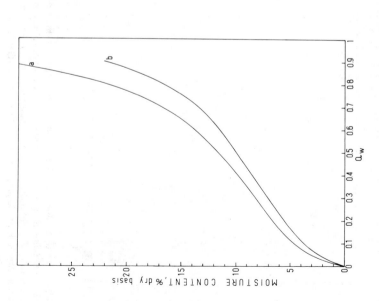

Fig. 161. Egg white (adsorption, 20°C): (a) egg white, (b) de-salted egg white. [Nemitz (1961), as quoted by Kuprianoff (1962).]

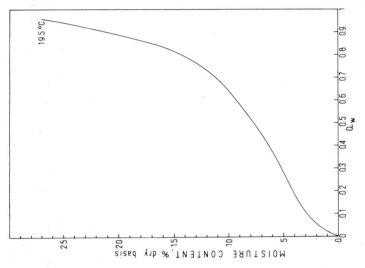

Fig. 164. Egg, whole (desorption, 195°C): Freeze-dried sample ground and allowed to equilibrate to about 20% (dry basis) moisture content before isotherm determination. Method: Manometric apparatus (Taylor, 1961.)

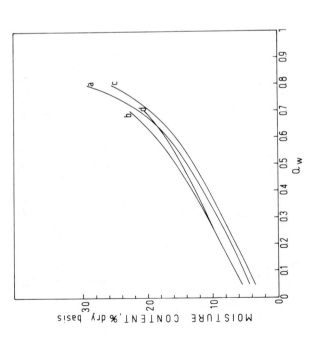

Fig. 163. Egg white (adsorption and desorption, 10°C): Freeze-dried at a maximum product temperature of 30°C; (a) raw, adsorption, (b) raw, desorption, (c) cooked, adsorption, (d) cooked, desorption. Method: Static-jar (sulfuric acid solutions). (Gane, 1943.)

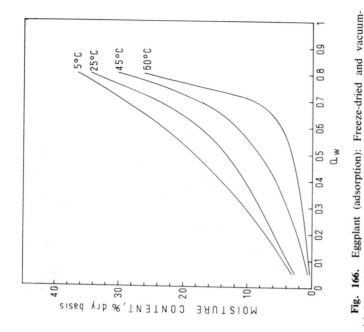

Fig. 166. Eggplant (adsorption): Freeze-dried and vacuum-dried at 30°C before adsorption; (a) 5, (b) 25, (c) 45, (d) 60°C. Method: Jar with air agitation (sulfuric acid solutions). (Wolf *et al.*, 1973.)

Fig. 165. Egg yolk (adsorption and desorption, 10°C): Freeze-dried at a maximum product temperature of 30°C; (a) raw, adsorption, (b) raw, desorption, (c) cooked, adsorption, (d) cooked, desorption. Method: Static-jar (sulfuric acid solutions). (Gane, 1943.)

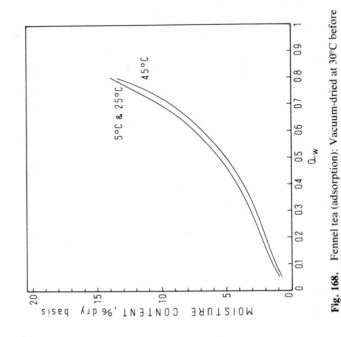

Fig. 168. Fennel tea (adsorption): Vacuum-dried at 30°C before adsorption; (a) 5, 25, (b) 45°C. Method: Jar with air agitation (sulfuric acid solutions). (Wolf *et al.*, 1973.)

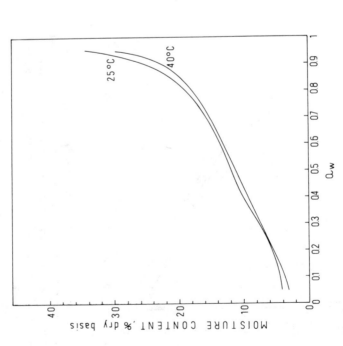

Fig. 167. Elastin (adsorption): Prepared from the ligamenta nuchae of beef by extraction with 0.9% NaCl and then with alcohol and ether, and then electrodialyzed. Method: Static-desiccator (sulfuric acid solutions). (Bull, 1944.)

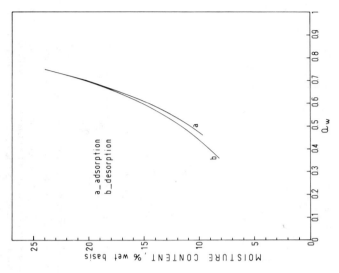

Fig. 170. Figs (adsorption and desorption, 25°C): Turkish "Lerida" harvested in 1972; samples for adsorption were air-dried at 30°C; samples for desorption were humidified with liquid water to 25% (wet basis) moisture content. Method: Dew-point. (Pixton and Warburton, 1976.)

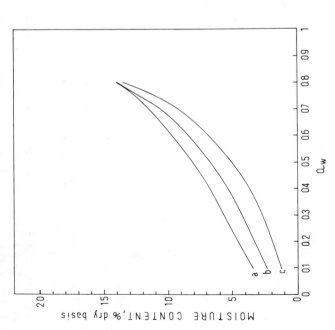

Fig. 169. Fennel tea (desorption): Vacuum-dried before adsorption; desorption following equilibration at 90% relative humidity; (a) 5, (b) 25, (c) 45°C. Method: Jar with air agitation (sulfuric acid solutions). (Wolf et al., 1973.)

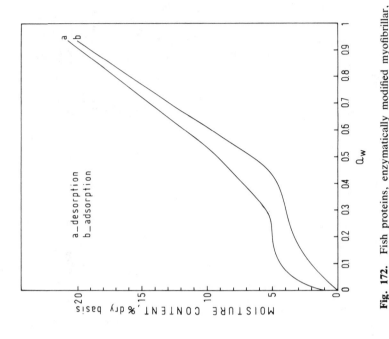

Fig. 172. Fish proteins, enzymatically modified myofibrillar, co-dried with 10% glucose (adsorption and desorption, 20°C): Fish protein isolates were prepared from partially hydrolyzed myofibrillar rockfish protein and co-dried (freeze-dried with an aqueous glucose solution). Method: Static-desiccator (saturated salt solutions). (Koury and Spinelli, 1975.)

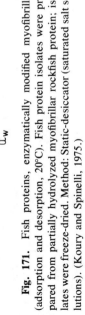

Fig. 171. Fish proteins, enzymatically modified myofibrillar (adsorption and desorption, 20°C). Fish protein isolates were prepared from partially hydrolyzed myofibrillar rockfish protein; isolates were freeze-dried. Method: Static-desiccator (saturated salt solutions). (Koury and Spinelli, 1975.)

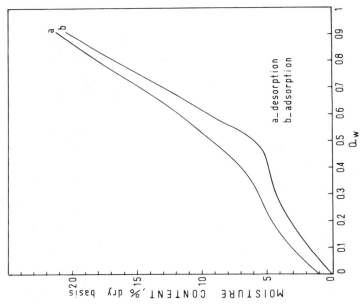

Fig. 173. Fish proteins, enzymatically modified myofibrillar, co-dried with 10% sucrose (adsorption and desorption, 20°C): Fish protein isolates were prepared from partially hydrolyzed myofibrillar rockfish protein and co-dried (freeze-dried) with an aqueous sucrose solution. Method: Static-desiccator (saturated salt solutions). (Koury and Spinelli, 1975.)

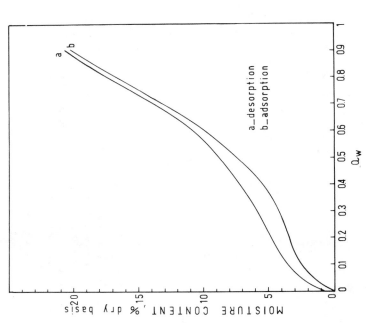

Fig. 174. Fish proteins, enzymatically modified myofibrillar, co-dried with 10% sorbitol (adsorption and desorption, 20°C): Fish protein isolates were prepared from partially hydrolyzed myofibrillar rockfish protein and co-dried (freeze-dried) with an aqueous sorbitol solution. Method: Static-desiccator (saturated salt solutions). (Koury and Spinelli, 1975.)

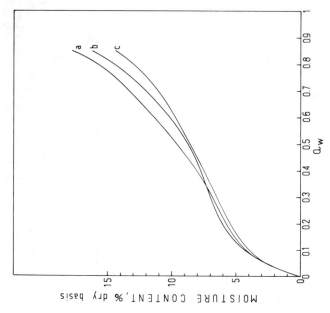

Fig. 176. Fish protein concentrate (adsorption, 25°C): (a) Non-steamed hake FPC milled by a Rietz disintegrator; (b) fine-particle size of steamed hake concentrate; (c) steamed menhaden concentrate milled by Rietz disintegrator. Method: Static-jar (saturated salt solutions). (Rasekh *et al.*, 1971.)

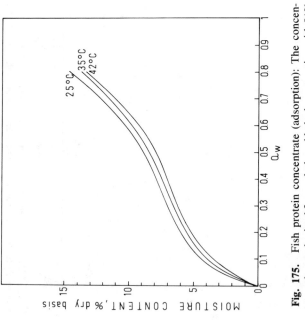

Fig. 175. Fish protein concentrate (adsorption): The concentrate used was obtained from whole red hake by extraction with 91% isopropyl alcohol followed by vacuum drying; the sample was milled by a Rietz desintegrator and then steam stripped at 105°C: (a) 25, (b) 35, (c) 42°C. Method: Static-jar (saturated salt solutions). (Rasekh *et al.*, 1971.)

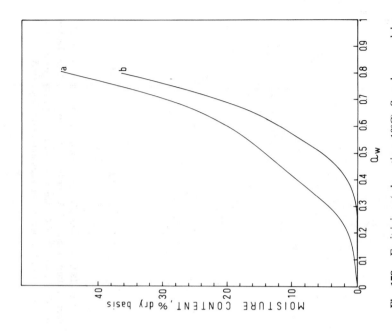

Fig. 178. Fruit juices (adsorption, 10°C): Samples were laboratory freeze-dried before adsorption; (a) black currant juice, (b) orange juice. Method: Static-jar (sulfuric acid solutions). (Gane, 1950.)

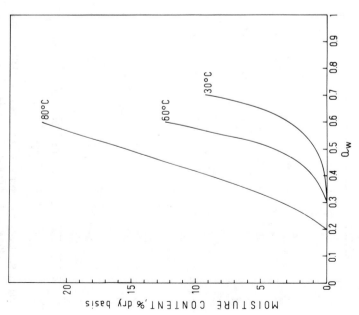

Fig. 177. Fructose: (a) 80, (b) 60, (c) 30°C. Method: Dynamic. (Audu *et al.*, 1978.)

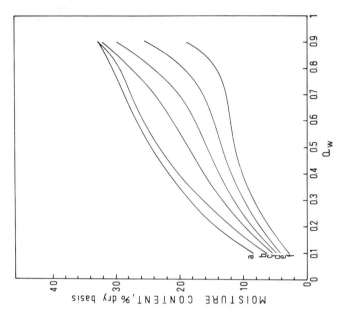

Fig. 180. Gelatin gel (adsorption): Freeze-dried at 30°C and then vacuum dried at 50°C for 6 hr; (a) −20, −10, 0, (b) 10, (c) 20, (d) 30, (e) 40, (f) 50°C. Method: Vacuum sorption apparatus with quartz spring balance. (Saravacos and Stinchfield, 1965.)

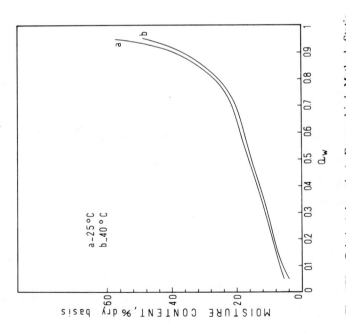

Fig. 179. Gelatin (adsorption): Freeze-dried. Method: Static-desiccator (sulfuric acid solutions). (Bull, 1944.)

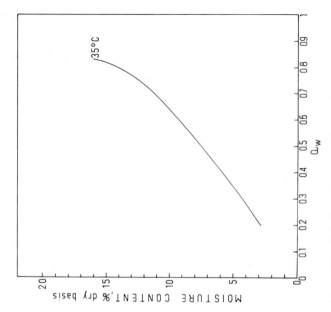

Fig. 182. Gelatin—microcrystalline cellulose (adsorption, 35°C): Gelatin and microcrystalline cellulose ("Avicel") in equal parts (1:1 dry basis) were wet-mixed and freeze-dried at room temperature before adsorption. Method: Static-desiccator (saturated salt/sulfuric acid solutions). (Iglesias *et al.*, 1980.)

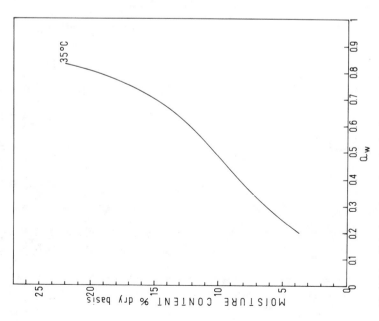

Fig. 181. Gelatin (adsorption, 35°C): Freeze-dried at room temperature. Method: Static-desiccator (saturated salt/sulfuric acid solutions). (Iglesias *et al.*, 1980.)

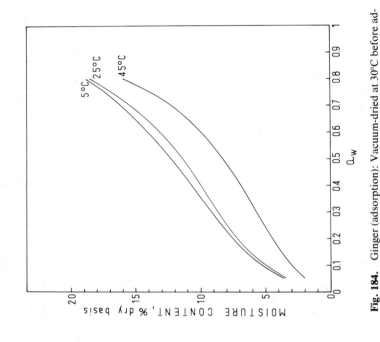

Fig. 183. Gelatin—starch gel [(a) desorption, (b) adsorption]: Gelatin and starch in equal parts (1:1 dry basis) were wet-mixed with continuous heating to produce starch gelatinization, and then freeze-dried at room temperature. Method: Static-desiccator (saturated salt/sulfuric acid solutions). (Iglesias *et al.*, 1980.)

Fig. 184. Ginger (adsorption): Vacuum-dried at 30°C before adsorption; (a) 5, (b) 25, (c) 45°C. Method: Jar with air agitation (sulfuric acid solutions). (Wolf *et al.*, 1973.)

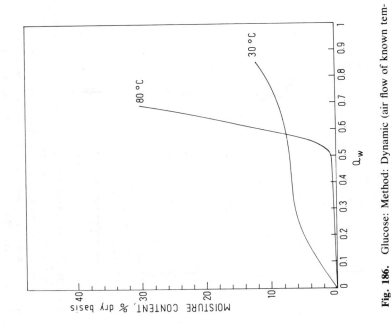

Fig. 186. Glucose: Method: Dynamic (air flow of known temperature and relative humidity). (Loncin *et al.*, 1968.)

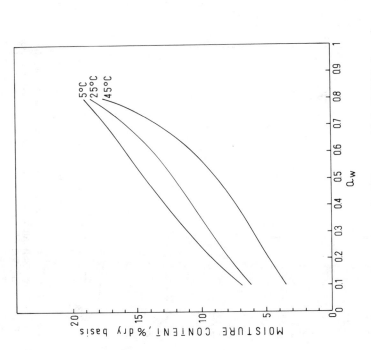

Fig. 185. Ginger (desorption): Vacuum-dried at 30°C before adsorption; desorption following humidification at 90% relative humidity; (a) 5, (b) 25, (c) 45°C. Method: Jar with agitation (sulfuric acid solutions). (Wolf *et al.*, 1973.)

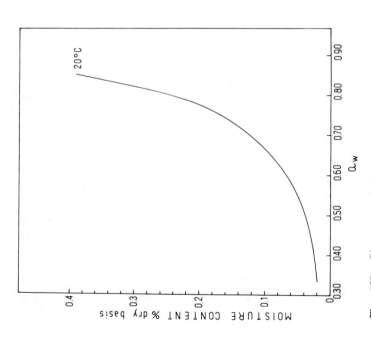

Fig. 188. Gluten, wheat (adsorption): Gluten was separated from an unbleached flour sample commercially milled from high-grade Canadian hard red spring wheat; the separated gluten was freeze-dried before adsorption and had a protein content of 69.8%; (a) 20.2, (b) 30.1, (c) 40.8, (d) 50.2°C. Method: Vacuum sorption apparatus with quartz spring balance. (Bushuk and Winkler, 1957.)

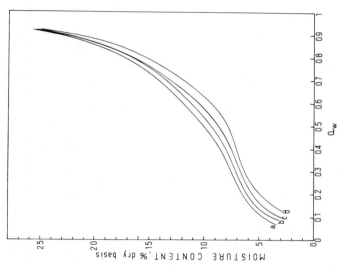

Fig. 187. Glutamate, sodium (adsorption, 20°C). Method: Static-desiccator (saturated salt/sulfuric acid solutions). (Lewicki and Brzozowski, 1973.)

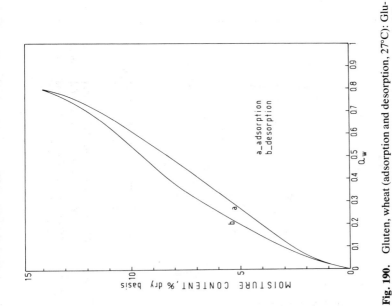

Fig. 189. Gluten, wheat (adsorption and desorption, 27°C): Gluten was separated from an unbleached flour sample commercially milled from high-grade Canadian hard red spring wheat; the separated gluten was freeze-dried and had a protein content of 69.8%. Method: Vacuum sorption apparatus with quartz spring balance. (Bushuk and Winkler, 1957.)

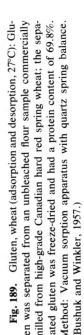

Fig. 190. Gluten, wheat (adsorption and desorption, 27°C): Gluten was separated from an unbleached flour sample commercially milled from high-grade Canadian hard red spring wheat; the separated gluten was spray-dried and had a protein content of 86.6%. Method: Vacuum sorption apparatus with quartz spring balance. (Bushuk and Winkler, 1957.)

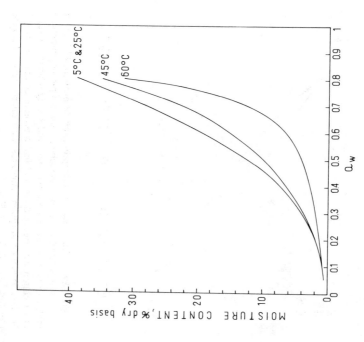

Fig. 192. Grapefruit (desorption, 5°C): Freeze-dried and vacuum-dried at 30°C before adsorption; desorption following humidification at 90% relative humidity. Method: Jar with air agitation (sulfuric acid solutions). (Wolf *et al.*, 1973.)

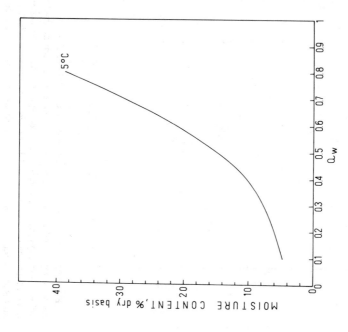

Fig. 191. Grapefruit (adsorption): Freeze-dried and vacuum-dried at 30°C before adsorption; (a) 5, 25, (b) 45, (c) 60°C. Method: Jar with air agitation (sulfuric acid solutions). (Wolf *et al.*, 1973.)

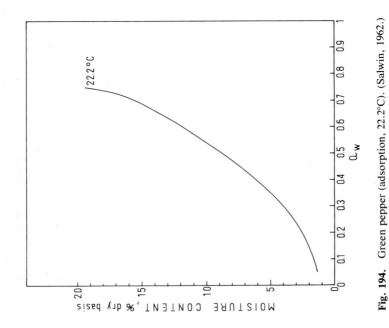

Fig. 194. Green pepper (adsorption, 22.2°C). (Salwin, 1962.)

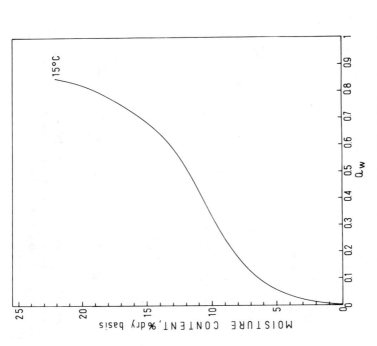

Fig. 193. Grass seed (sorption, 5°C): Initial water content 10.3%; grass seed: Chewing's Fescue 1938 crop from Garton's Ltd., Warrington, England. Method: Static-jar (sulfuric acid solutions). (Gane, 1948.)

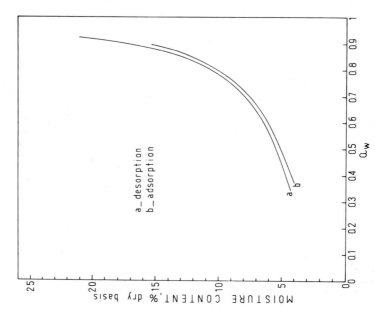

Fig. 196. Groundnut, kernels (adsorption and desorption, 20°C): Adsorption: commercial sample from Nigeria; desorption: commercial sample purchased in England. Method: Dew-point. (Ayerst, 1965.)

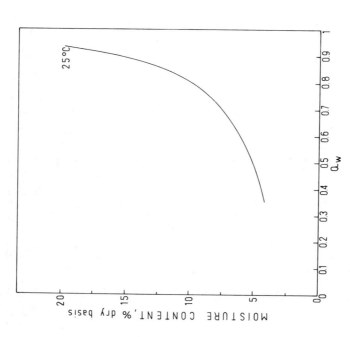

Fig. 195. Groundnuts (adsorption, 25°C): Samples were air-dried at 30°C until $a_w \approx 0.15$–0.20 prior to adsorption; mean oil content 50.6% (dry basis). Method: Dew-point. (Pixton and Warburton, 1971b.)

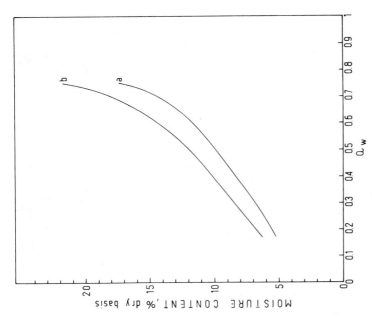

Fig. 198. Guava-Taro (sorption, 22°C): Commercial whole white taro (Apii variety) corms and guava (Beaumont variety) puree were combined for drum drying; guava:taro flakes ratio, (a) 1 : 2, (b) 3 : 2. Method: Static-desiccator (saturated salt solutions). (Nip, 1978.)

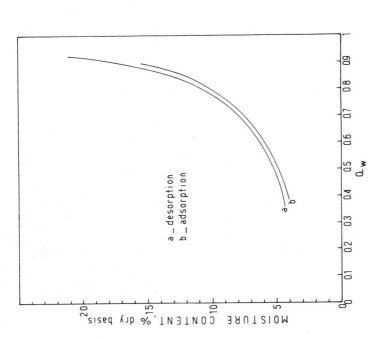

a _ desorption
b _ adsorption

Fig. 197. Groundnut, Kernels (adsorption and desorption, 30°C): Adsorption: commercial sample from Nigeria; desorption: commercial sample purchased in England. Method: Dew-point. (Ayerst, 1965.)

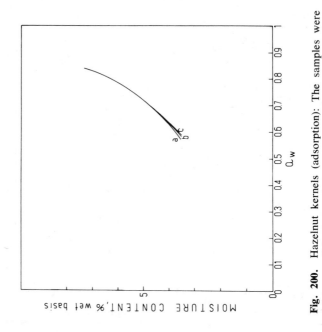

Fig. 200. Hazelnut kernels (adsorption): The samples were grated on a small food cutter before determination of adsorption isotherm: (a) 20, (b) 40, (c) 30°C. Method: Dew-point. (Ayerst, 1965.)

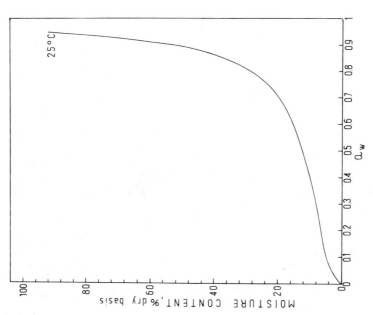

Fig. 199. Halibut (adsorption, 25°C): The muscle of halibut (*Hippoglossus stenolepis*) was freeze-dried and ground into coarse powder. Method: Static-desiccator (sulfuric acid solutions). (Koizumi et al., 1978.)

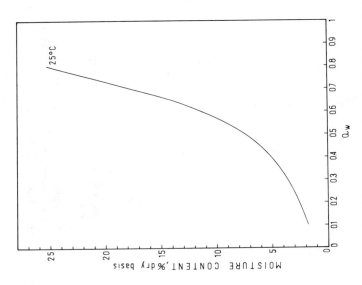

Fig. 201. Hibiscus tea (adsorption, 25°C): Vacuum-dried at 30°C before adsorption. Method: Jar with air agitation (sulfuric acid solutions). (Wolf *et al.*, 1973.)

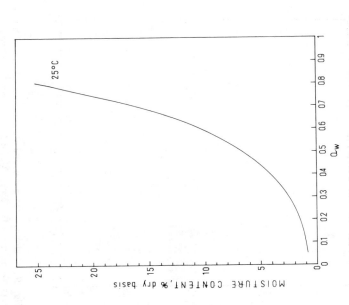

Fig. 202. Hibiscus tea (desorption, 25°C): Vacuum-dried at 30°C before adsorption; desorption following humidification at 90% relative humidity. Method: Jar with air agitation (sulfuric acid solutions). (Wolf *et al.*, 1973.)

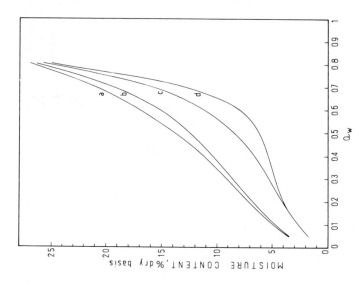

Fig. 204. Horseradish (adsorption): Freeze-dried and vacuum-dried at 30°C before adsorption: (a) 5, (b) 25, (c) 45, (d) 60°C. Method: Jar with air agitation (sulfuric acid solutions). (Wolf *et al.*, 1973.)

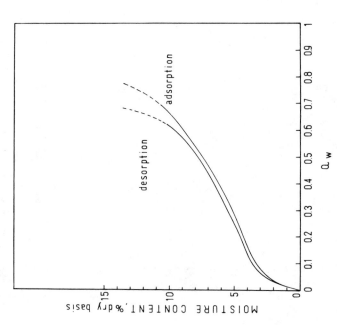

Fig. 203. Hops (adsorption and desorption, 25°C) California clusters from the 1971 crop. Method: Dew-point. (Henderson, 1973.)

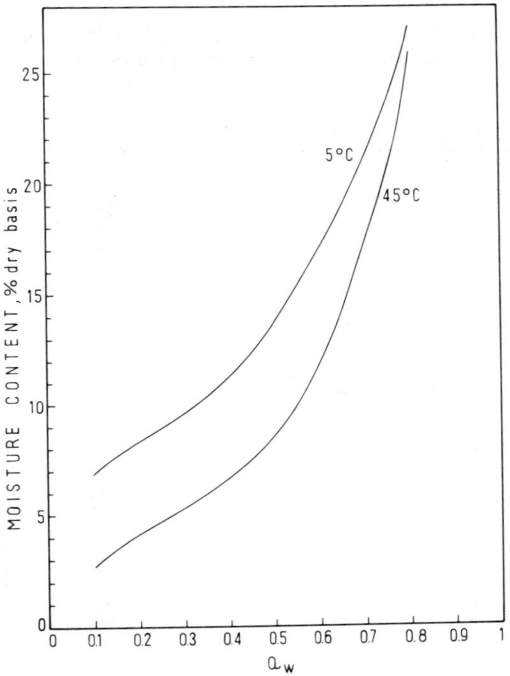

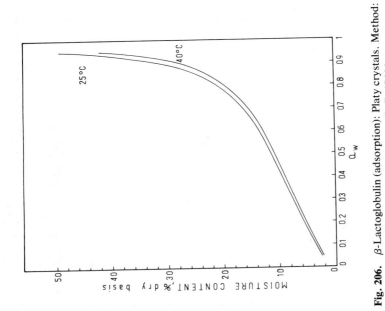

Fig. 205. Horseradish (desorption): Freeze-dried and vacuum-dried at 30°C before adsorption; desorption following humidification at 90% relative humidity. Method: Jar with air agitation (sulfuric acid solutions). (Wolf *et al.*, 1973.)

Fig. 206. β-Lactoglobulin (adsorption): Platy crystals. Method: Static-desiccator (sulfuric acid solutions). (Bull, 1944.)

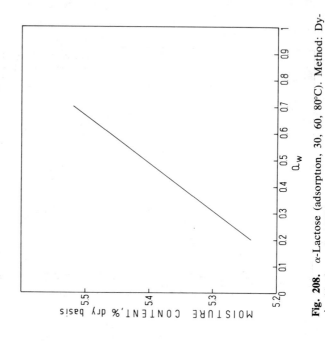

Fig. 208. α-Lactose (adsorption, 30, 60, 80°C). Method: Dynamic. (Audu *et al.*, 1978.)

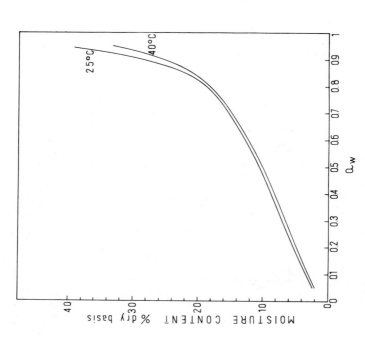

Fig. 207. β-Lactoglobulin (adsorption): Platy crystals, freeze-dried. Method: Static-desiccator (sulfuric acid solutions). (Bull, 1944.)

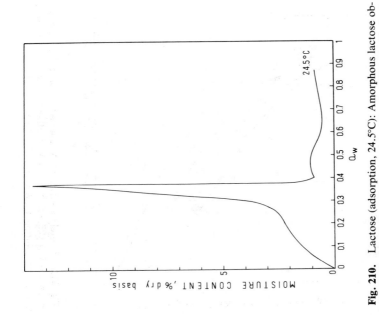

Fig. 210. Lactose (adsorption, 24.5°C): Amorphous lactose obtained by freeze-drying of an aqueous lactose solution. Method: Electrobalance assembly. (Berlin *et al.*, 1968b.)

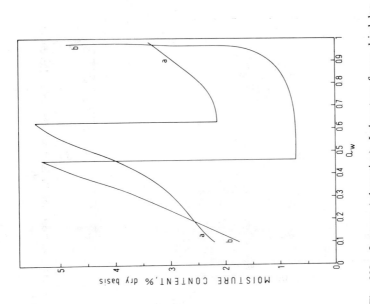

Fig. 209. Lactose (adsorption): Laboratory freeze-dried lactose: (a) 14, (b) 34°C. Method: Electrobalance assembly. (Berlin *et al.*, 1970.)

115

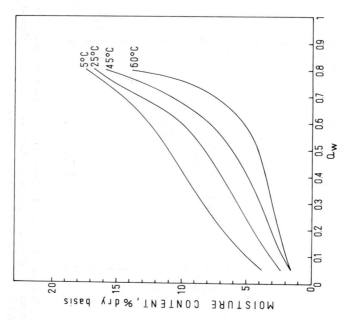

Fig. 212. Laurel (adsorption): Vacuum-dried at 30°C before adsorption; (a) 5, (b) 25, (c) 45, (d) 60°C. Method: Jar with air agitation (sulfuric acid solutions). (Wolf *et al.*, 1973.)

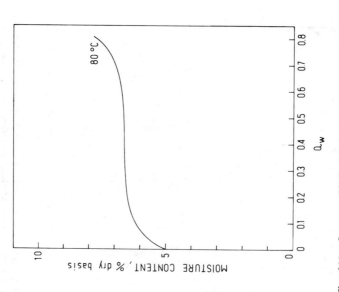

Fig. 211. Lactose (80°C). Method: Dynamic. (Loncin *et al.*, 1968.)

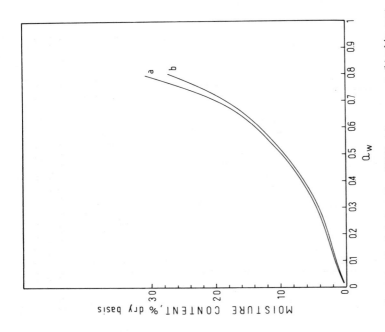

Fig. 214. Leek (adsorption, 22°C): (a) green part, (b) white part. Method: Static-desiccator (saturated salt/sulfuric acid solutions). (Lewicki and Lenart, 1975.)

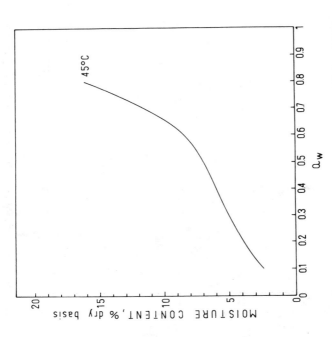

Fig. 213. Laurel (desorption, 45°C): Vacuum-dried at 30°C before adsorption: desorption following humidification at 90% relative humidity. Method: Jar with air agitation (sulfuric acid solutions). (Wolf *et al.*, 1973.)

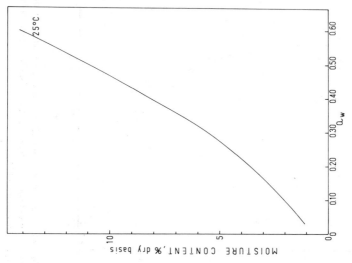

Fig. 216. Lemon crystals (adsorption, 25°C): Lemon juice (30°Bx) was freeze-dried with radiation plates kept at 45°C; the product was further dried in a vacuum oven at 15°C. Method: Electrobalance assembly. (Kopelman *et al.*, 1977.)

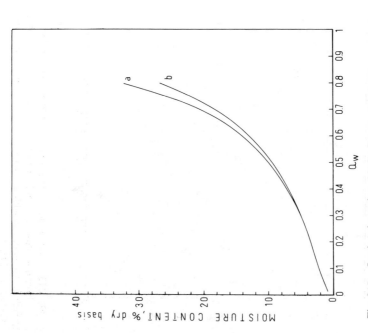

Fig. 215. Leek (adsorption, 22°C): Freeze-dried: (a) green part, (b) white part. Method: Static-desiccator (saturated salt/sulfuric acid solutions). (Lewicki and Lenart, 1975.)

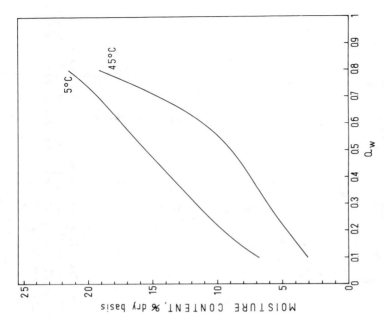

Fig. 218. Lentil (desorption): Vacuum-dried at 30°C before adsorption; desorption following humidification at 90% relative humidity. Method: Jar with air agitation (sulfuric acid solutions). (Wolf *et al.*, 1973.)

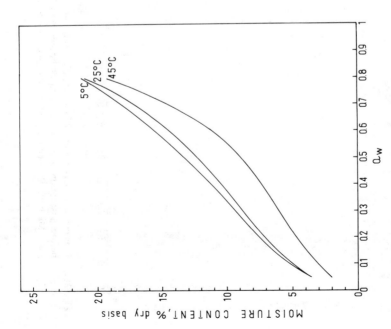

Fig. 217. Lentil (adsorption): Vacuum-dried at 30°C before adsorption: (a) 5, (b) 25, (c) 45°C. Method: Jar with air agitation (sulfuric acid solutions). (Wolf *et al.*, 1973.)

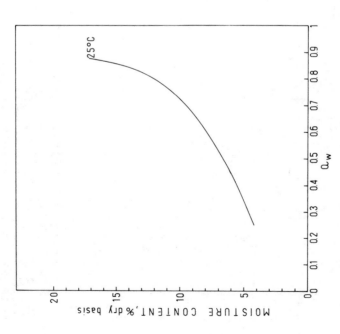

Fig. 220. Linseed seed (sorption, 15°C): Linseed was Dutch H 1935 crop from Plant Breeding Station, Stormont, Belfast; initial moisture content was 7.3% (dry basis). Method: Static-jar (sulfuric acid solutions). (Gane, 1948.)

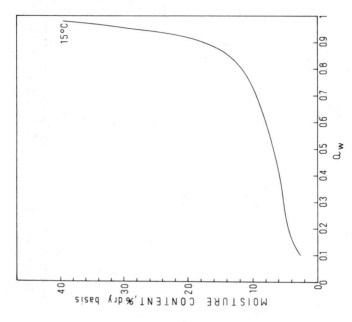

Fig. 219. Linseed seed (adsorption, 25°C): Samples were air-dried at 30°C before adsorption; mean oil content of linseed was 40.9% (dry basis). Method: Dew-point. (Pixton and Warburton, 1971b.)

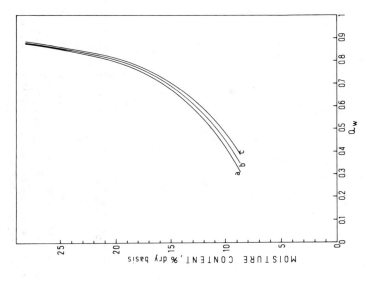

Fig. 222. Lupins, bitter blue (desorption): Oil content was 5.8% (dry basis); desorption data were obtained by adding liquid water to raise a_w to 0.95; (a) 15, (b) 25, (c) 35°C. Method: Dew-point. (Pixton and Henderson, 1979.)

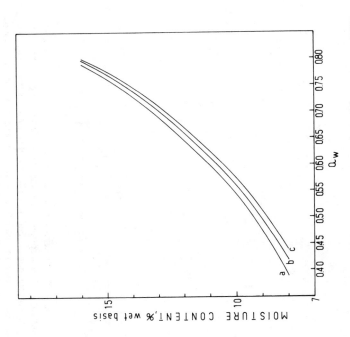

Fig. 221. Lupins, bitter blue (adsorption): Oil content was 5.8% (dry basis); adsorption data were obtained from samples air-dried at 30°C; (a) 15, (b) 25, (c) 35°C. Method: Dew-point. (Pixton and Henderson, 1979.)

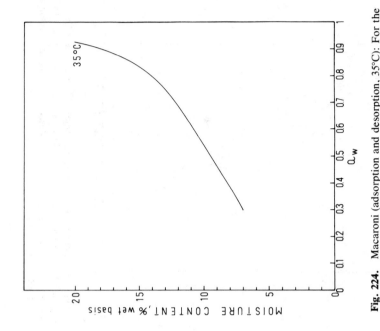

Fig. 224. Macaroni (adsorption and desorption, 35°C): For the determination of adsorption isotherms the samples were air-dried at a temperature not exceeding 35°C; for desorption liquid water was added to raise the moisture content of macaroni. Method: Dew-point. (Pixton and Warburton, 1973b.)

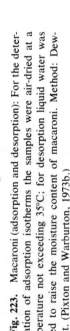

Fig. 223. Macaroni (adsorption and desorption): For the determination of adsorption isotherms the samples were air-dried at a temperature not exceeding 35°C; for desorption liquid water was added to raise the moisture content of macaroni. Method: Dew-point. (Pixton and Warburton, 1973b.)

122

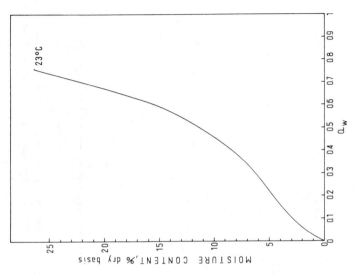

Fig. 225. Macaroni (adsorption, 25°C): (a) Wetting previously dried macaroni with liquid water; (b) wetting macaroni as received by exposure to atmosphere of controlled humidities. Method: Dew-point. (Pixton and Warburton, 1973b.)

Fig. 226. Maltose (adsorption, 23°C): 18.8% (w/w) aqueous maltose solution freeze-dried at ambient temperature. Method: Static-desiccator (saturated salt solutions). (Flink and Karel, 1972.)

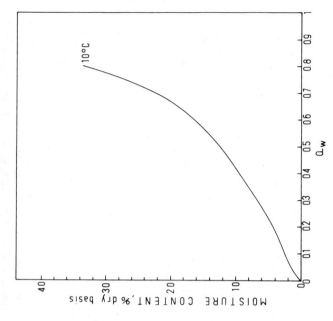

Fig. 228. Marrow (adsorption, 10°C): Scalded and laboratory freeze-dried. Method: Static-jar (sulfuric acid solutions). (Gane 1950.)

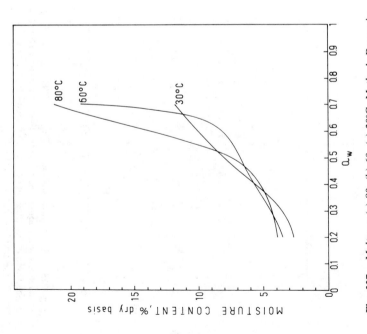

Fig. 227. Maltose: (a) 80, (b) 60, (c) 30°C. Method: Dynamic. (Audu *et al.*, 1978.)

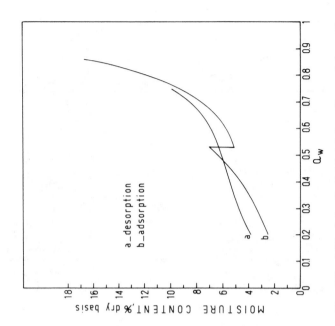

Fig. 229. Milk, baby food dried (adsorption and desorption, 25°C): The product was made almost entirely from standardized buffalo milk, roller-dried at 139.4°C; composition of the powder (wet basis): fat, 18%; protein, 22%; lactose, 33%; sucrose, 19%; minerals and vitamins, 4.5%. Method: Dynamic with continuous weighing. (Varshney and Ojha, 1977.)

Fig. 230. Milk, baby food dried (adsorption and desorption, 25°C): The product was made almost entirely from standardized buffalo milk spray-dried at 210°C inlet air temperature; composition of the powder (wet basis): fat, 18%; protein, 26%; lactose, 33%; sucrose, 14%; minerals and vitamins, 6%. Method: Dynamic with continuous weighing. (Varshney and Ojha, 1977.)

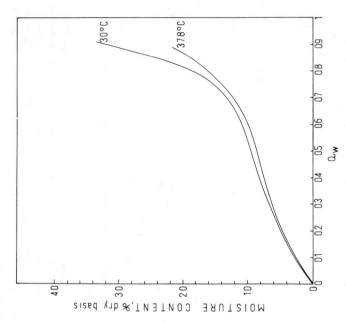

Fig. 232. Milk, nonfat dry (adsorption): Nonfat spray-dried milk was exposed at 61.6°C for 30 min prior to adsorption (low heat). Method: Static-sealed container (saturated salt solutions). (Heldman *et al.*, 1965.)

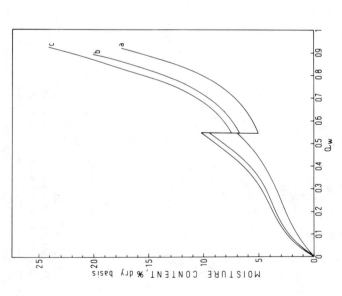

Fig. 231. Milk (adsorption, 24.5°C): Spray-dried in a laboratory pilot plant; (a) whole milk; (b) whole milk data corrected to a solid nonfat basis; (c) skim milk. Method: Electrobalance assembly. (Berlin *et al.*, 1968a.)

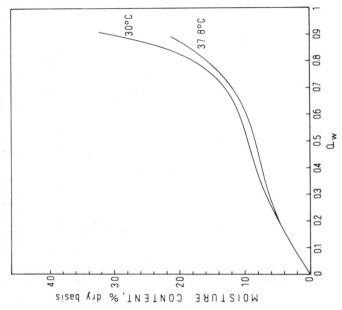

Fig. 234. Milk, nonfat dry (adsorption): Nonfat spray-dried milk was exposed at 73.8°C for 25 min prior to isotherm determination (medium heat). Method: Static-sealed container (saturated salt solutions). (Heldman *et al.*, 1965.)

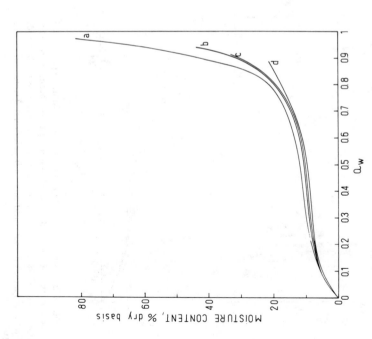

Fig. 233. Milk, nonfat dry (desorption): Nonfat spray-dried milk was exposed at 61.6°C for 30 min prior to isotherm determination (low heat); (a) 1.7, (b) 15.5, (c) 30, (d) 37.8°C. Method: Static-sealed container (saturated salt solutions). (Heldman *et al.*, 1965.)

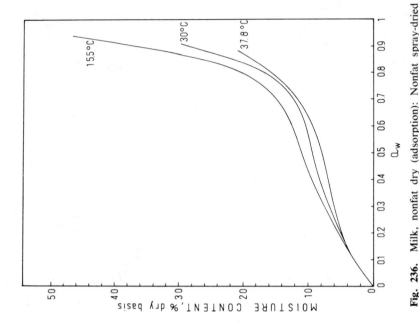

Fig. 235. Milk, nonfat dry (desorption): Nonfat spray-dried milk was exposed at 73.8°C for 25 min prior to isotherm determination (medium heat); (a) 1.7, (b) 15.5, (c) 30, (d) 37.8°C. Method: Static-sealed container (saturated salt solutions). (Heldman *et al.*, 1965.)

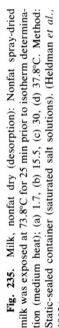

Fig. 236. Milk, nonfat dry (adsorption): Nonfat spray-dried milk was exposed at 85°C for 20 min prior to isotherm determination (high heat); (a) 15.5, (b) 30, (c) 37.8°. Method: Static-sealed container (saturated salt solutions). (Heldman *et al.*, 1965.)

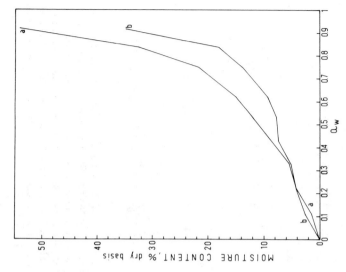

Fig. 238. Milk, lactose hydrolyzed nonfat (adsorption, 25°C): The nonfat milk (0.05% fat) was pasteurized and lactose hydrolysis was done with Maxilact 40,000® (purified β-galactosidase from *Saccharomyces lactis*); milk was then spray-dried with 170°C inlet air temperature and used for isotherm determination; (a) 50% lactose hydrolyzed; (b) not hydrolyzed. Method: Static-desiccator (saturated salt solutions) and electrobalance assembly. (San José *et al.*, 1977.)

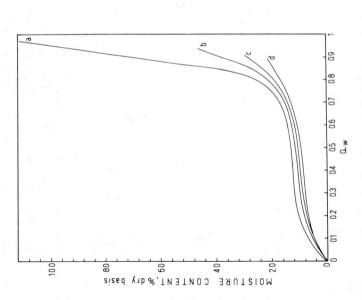

Fig. 237. Milk, nonfat dry (desorption): Nonfat spray-dried milk was exposed at 85°C for 20 min prior to isotherm determination (high heat); (a) 1.7, (b) 15.5, (c) 30, (d) 37.8°C. Method: Static-sealed container (saturated salt solutions). (Heldman *et al.*, 1965.)

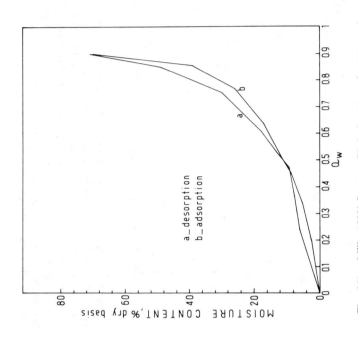

Fig. 240. Milk, 100% Lactose Hydrolyzed nonfat (adsorption and desorption): Nonfat milk (0.05% fat) was pasteurized and subjected to lactose hydrolysis by incubation with Maxilact 40,000® (purified β-galactosidase from *S. lactis*); the milk was then spray-dried and used for isotherm determination. Method: Static-desiccator (saturated salt solutions) and electrobalance assembly. (San José *et al.*, 1977.)

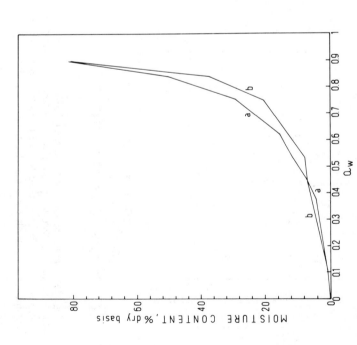

Fig. 239. Milk, 100% lactose hydrolyzed low-fat [(a) desorption and (b) adsorption, 25°C]: Low-fat milk (0.5% fat) was UHT treated and lactose hydrolysis was done by incubation with Maxilact 40,000® (purified β-galactosidase from *S. lactis*); the milk was then freeze-dried and used for isotherm determination. Method: Static-desiccator (saturated salt solutions) and electrobalance assembly. (San José *et al.*, 1977.)

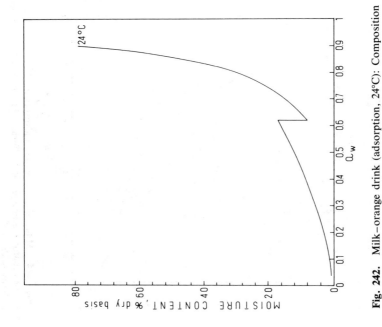

Fig. 241. Milk, 100% lactose hydrolyzed nonfat (adsorption and desorption): Nonfat milk (0.05% fat) was pasteurized and subjected to lactose hydrolysis by incubation with Maxilact 40,000® (purified β-galactosidase from *S. lactis*); the milk was then spray-dried (170°C inlet air temperature) and used for isotherm determination. Method: Static-desiccator (saturated salt solutions) and electrobalance assembly. (San José *et al.*, 1977.)

Fig. 242. Milk–orange drink (adsorption, 24°C): Composition of beverage (foam, spray-dried) (dry basis): whole milk solids, 41.4%; orange juice solids, 27.1%; sucrose, 30.3%; carboxymethyl cellulose, 1.2%. Method: Electrobalance assembly. (Berlin *et al.*, 1973b.)

131

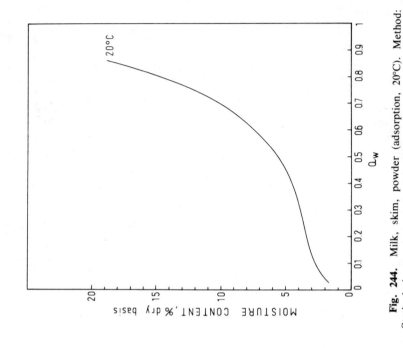

Fig. 243. Milk powder components (adsorption, 24.5°C): All components were freeze-dried prior to isotherm determination; (a) centrifugal casein: ash content, 7.49%; calcium content, 2.87%; (b) acid-precipitated casein: ash content, 7.74%; calcium content, 0.15%; (c) acid whey. Method: Electrobalance assembly. (Berlin *et al.,* 1968b.)

Fig. 244. Milk, skim, powder (adsorption, 20°C). Method: Static-desiccator (saturated salt/sulfuric acid solutions). (Lewicki and Brzozowski, 1973.)

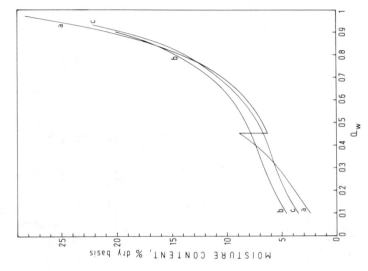

Fig. 245. Milk, skim (adsorption): Freeze-dried in a laboratory pilot plant: (a) 14, (b) 24, (c) 34°C. Method: Electrobalance assembly. (Berlin *et al.*, 1970.)

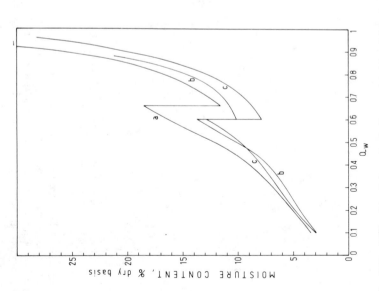

Fig. 246. Milk, skim (adsorption and desorption, 34°C): The data show successive adsorption–desorption cycles on spray-dried (laboratory pilot plant) skim milk powder; (a) adsorption, first cycle; (b) desorption, first cycle; (c) adsorption, second cycle. Method: Electrobalance assembly. (Berlin *et al.*, 1970.)

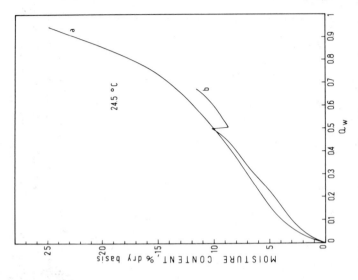

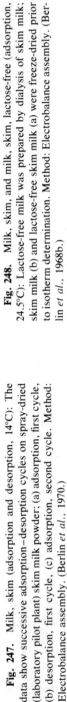

Fig. 248. Milk, skim, and milk, skim, lactose-free (adsorption, 24.5°C): Lactose-free milk was prepared by dialysis of skim milk; skim milk (b) and lactose-free skim milk (a) were freeze-dried prior to isotherm determination. Method: Electrobalance assembly. (Berlin *et al.*, 1968b.)

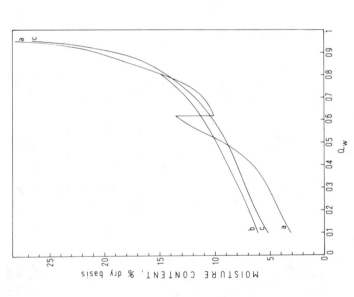

Fig. 247. Milk, skim (adsorption and desorption, 14°C): The data show successive adsorption–desorption cycles on spray-dried (laboratory pilot plant) skim milk powder; (a) adsorption, first cycle, (b) desorption, first cycle, (c) adsorption, second cycle. Method: Electrobalance assembly. (Berlin *et al.*, 1970.)

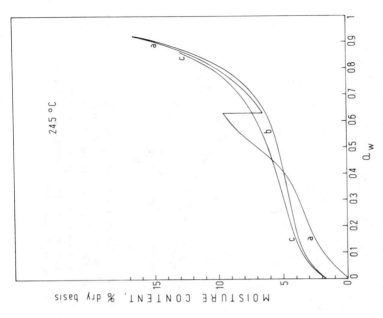

Fig. 250. Milk, whole [(a) adsorption, (b) desorption, (c) read-sorption]. Adsorption was conducted on foam-spray-dried (laboratory pilot plant) whole milk. Method: Electrobalance assembly. (Berlin et al., 1968a.)

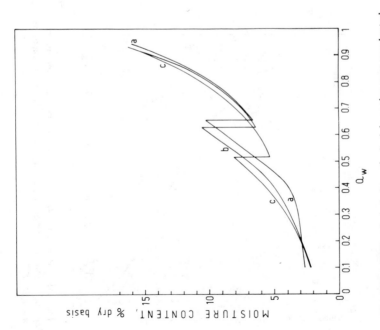

Fig. 249. Milk, whole (adsorption): Adsorption was conducted on foam-spray-dried (laboratory pilot plant) whole milk; (a) 14, (b) 24, (c) 34°C. Method: Electrobalance assembly. (Berlin *et al.*, 1970.)

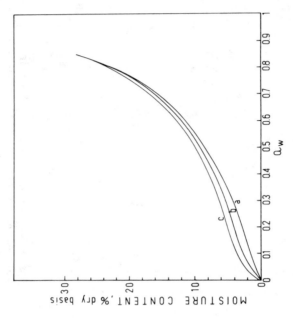

Fig. 252. Muscle fibers (adsorption): Muscle tissue samples were cut from the longissimus dorsi taken from beef carcasses and frozen at different freezing rates before freeze-drying; (a) pieces 2.5 cm thick in still air at −10°C, (b) 1-cm cubes in still air at −30°C, (c) slices 1 mm thick at −150°C. Method: Microbalance technique. (MacKenzie and Luyet, 1967.)

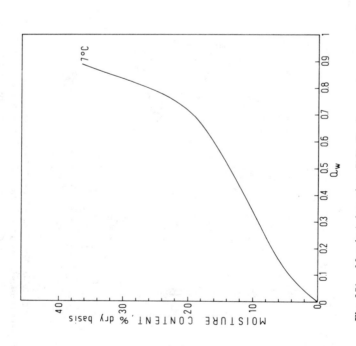

Fig. 251. Muscle (adsorption, 7°C): Bovine arm bone muscle. Method: Vacuum sorption apparatus with quartz spring balance. (Walker et al., 1973.)

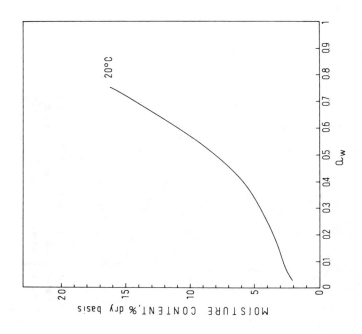

Fig. 253. Mushrooms (*A. bisporus*) (adsorption): Freeze-dried samples; (a) 25, (b) 20°C. Method: Static-desiccator (saturated salt solutions). [(a) Lafuente and Piñaga (1966), (b) Iglesias (1975).]

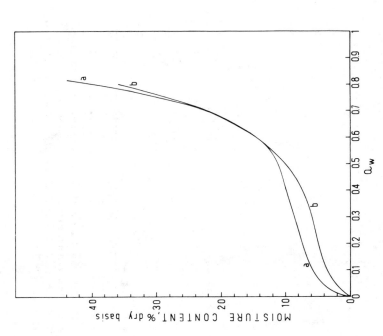

Fig. 254. Mushrooms, dried (adsorption, 20°C). Method: Static-desiccator (saturated salt/sulfuric acid solutions. (Lewicki and Brzozowski, 1973.)

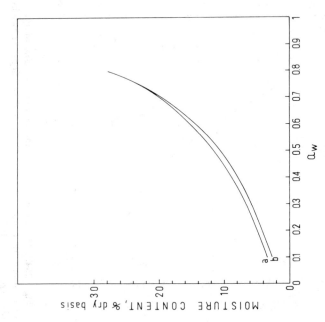

Fig. 256. Mushrooms (*Boletus*) (desorption): Dried samples were further vacuum-dried at 30°C before adsorption; desorption following humidification at 90% relative humidity; (a) 5, (b) 25°C. Method: Jar with air agitation (sulfuric acid solutions). (Wolf *et al.*, 1973.)

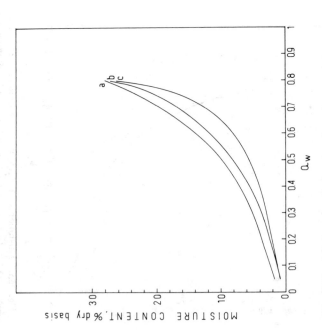

Fig. 255. Mushrooms (*Boletus*) (adsorption): Dried samples were further vacuum-dried at 30°C before adsorption; (a) 5, 25, (b) 45, (c) 60°C. Method: Jar with air agitation (sulfuric acid solutions). (Wolf *et al.*, 1973.)

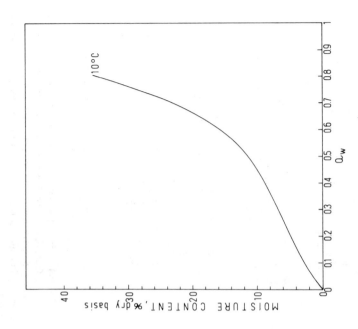

Fig. 258. Mushrooms (adsorption, 10°C): Samples were freeze-dried. Method: Static-jar (sulfuric acid solutions). (Gane, 1950.)

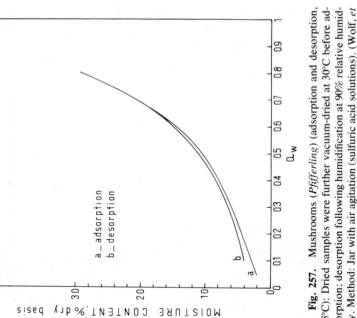

Fig. 257. Mushrooms (*Pfifferling*) (adsorption and desorption, 25°C): Dried samples were further vacuum-dried at 30°C before adsorption; desorption following humidification at 90% relative humidity. Method: Jar with air agitation (sulfuric acid solutions). (Wolf, *et al.*, 1973.)

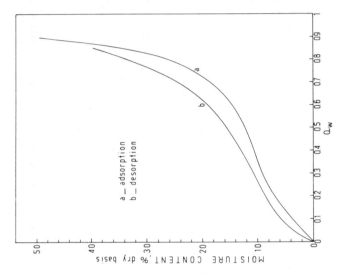

Fig. 260. Myosin A (adsorption and desorption, 7°C): Myosin A extracted from bovine arm bone muscle; after final isoelectric precipitation, the protein was dialyzed and freeze-dried. Method: Vacuum sorption apparatus with quartz spring balance. (Walker *et al.*, 1973.)

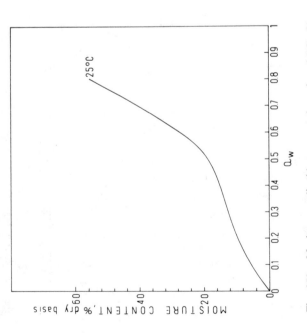

Fig. 259. Mushrooms, liquid extract (adsorption, 25°C): Var. *A. bisporus*. Static-desiccator (saturated salt solutions). (Bartholomai *et al.*, 1975.)

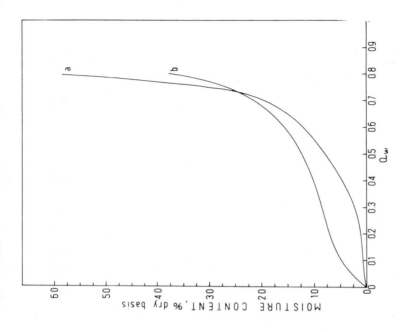

Fig. 262. Myosin B (adsorption, 7°C): Myosin B extracted from bovine arm bone muscle; (a) Nondialyzed, (b) dialyzed. Method: Vacuum sorption apparatus with quartz spring balance. (Walker *et al.*, 1973.)

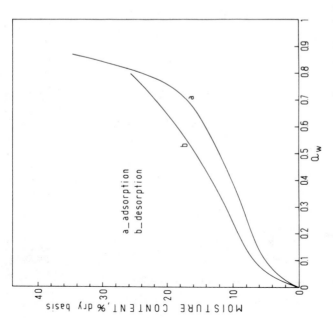

a_adsorption
b_desorption

Fig. 261. Myosin B (adsorption and desorption, 7°C): Myosin B extracted from bovine arm bone muscle; after final isoelectric precipitation the protein was dialyzed and freeze-dried. Method: Vacuum sorption apparatus with quartz spring balance. (Walker *et al.*, 1973.)

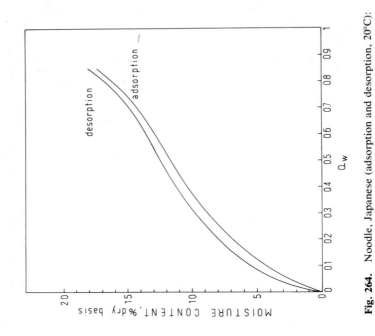

Fig. 264. Noodle, Japanese (adsorption and desorption, 20°C): Dried Japanese noodle (udon), with 0% (dry basis) NaCl. (Shibata *et al.*, 1976.)

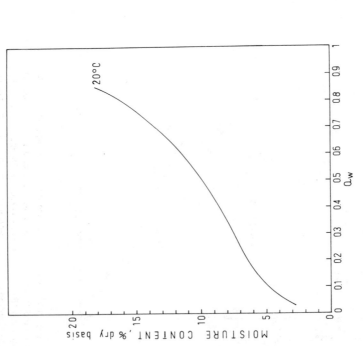

Fig. 263. Noodles (adsorption, 20°C). Method: Static-desiccator (saturated salt/sulfuric acid solutions). (Lewicki and Brzozowski, 1973.)

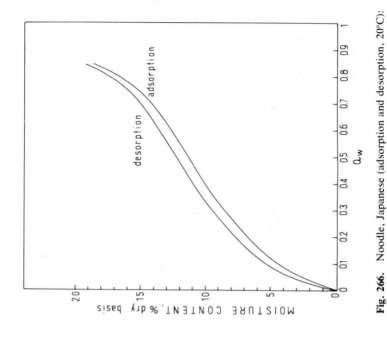

Fig. 265. Noodle, Japanese (adsorption and desorption, 20°C): Dried Japanese noodle (udon), with 1.14% (dry basis) NaCl. (Shibata *et al.*, 1976.)

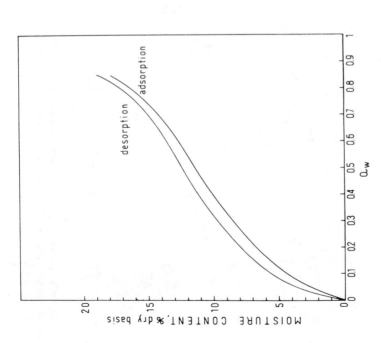

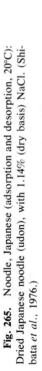

Fig. 266. Noodle, Japanese (adsorption and desorption, 20°C): Dried Japanese noodle (udon), with 2.23% (dry basis) NaCl. (Shibata *et al.*, 1976.

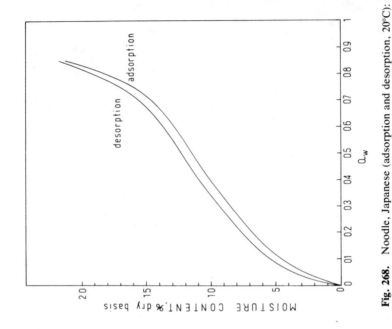

Fig. 268. Noodle, Japanese (adsorption and desorption, 20°C): Dried Japanese noodle (udon), with 4.37% (dry basis) NaCl. (Shibata *et al.*, 1976.)

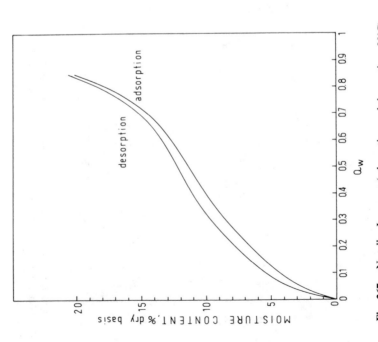

Fig. 267. Noodle, Japanese (adsorption and desorption, 20°C): Dried Japanese noodle (udon), with 3.30% (dry basis) NaCl. (Shibata *et al.*, 1976.)

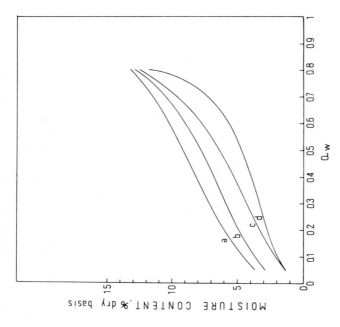

Fig. 270. Nutmeg (adsorption): Samples were vacuum-dried at 30°C before adsorption; (a) 5, (b) 25, (c) 45, (d) 60°C. Method: Jar with air agitation (sulfuric acid solutions). (Wolf et al., 1973.)

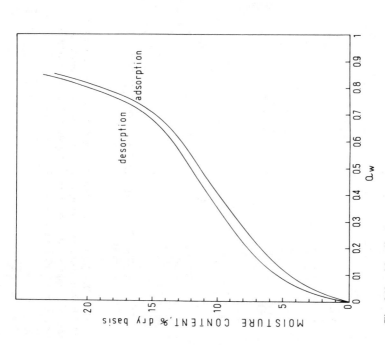

Fig. 269. Noodle, Japanese (adsorption and desorption, 20°C): Dried Japanese noodle (udon), with 5.39% (dry basis) NaCl. (Shibata et al., 1976.)

145

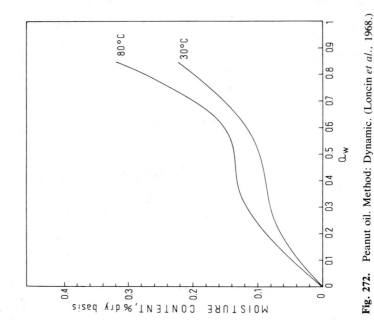

Fig. 272. Peanut oil. Method: Dynamic. (Loncin *et al.*, 1968.)

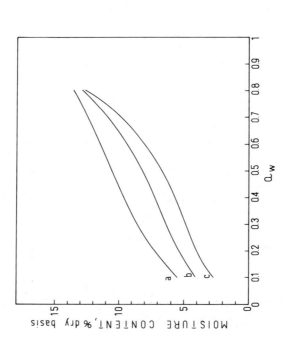

Fig. 271. Nutmeg (desorption): Samples were vacuum-dried at 30°C before adsorption; desorption following humidification at 90% relative humidity; (a) 5, (b) 25, (c) 45°C. Method: Jar with air agitation (sulfuric acid solutions). (Wolf *et al.*, 1973.)

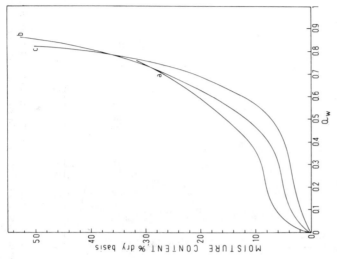

Fig. 274. Onion (adsorption): Dehydrated chopped onions (Yellow Globe, CV. Autumn Spice, Alberta, Canada), with moisture content 5.2%, were used; samples were air-dried at 90°C for 1 hr and then at 41°C for 15 hr; (a) 10, (b) 30, (c) 45°C. Method: Static-desiccator (saturated salt solutions). (Mazza and Le Maguer, 1978.)

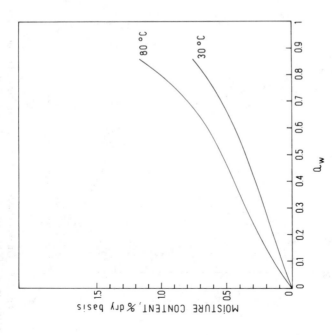

Fig. 273. Oleic acid. Method: Dynamic. (Loncin *et al.*, 1968.)

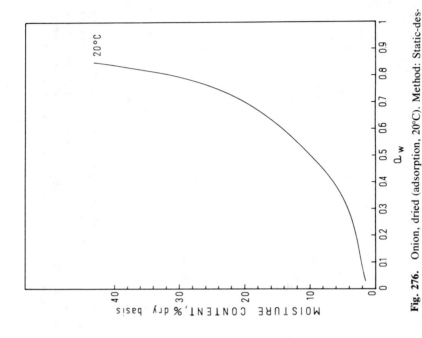

Fig. 276. Onion, dried (adsorption, 20°C). Method: Static-desiccator (saturated salt/sulfuric acid solutions). (Lewicki and Brzozowski, 1973.)

Fig. 275. Onion (desorption): Dehydrated chopped onions (Yellow Glove CV. Autumn Spice, Alberta, Canada), with moisture content 5.2%, were equilibrated at 100% relative humidity before desorption experiments; samples were air-dried at 90°C for 1 hr and then at 41°C for 15 hr; (a) 10, (b) 30, (c) 45°C. Method: Static-desiccator (saturated salt solutions). (Mazza and Le Maguer, 1978.)

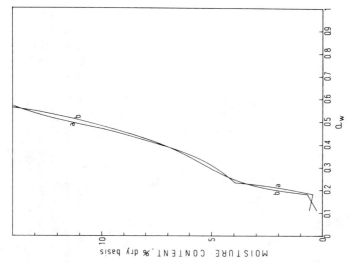

Fig. 278. Onion, powdered (adsorption, 25°C): Commercial dehydrated onion flakes ground and further vacuum-dried; conditioner used was calcium stearate (1%); (a) without, (b) with conditioner. Method: Static-desiccator (saturated salt/sulfuric acid solutions). (Peleg and Mannheim, 1977.)

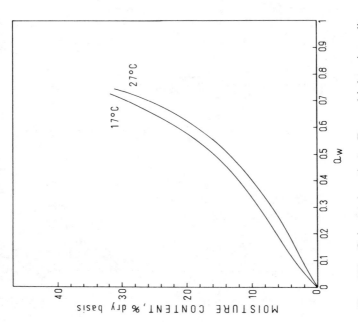

Fig. 277. Onion (adsorption): Freeze-dried onion slices. Method: Manometric apparatus. (Alcaraz et al., 1977.)

149

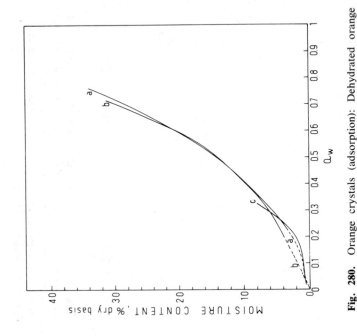

Fig. 280. Orange crystals (adsorption): Dehydrated orange juice with moisture content less than 0.5% was further dehydrated at room temperature to less than 0.1%: (a, c) 25, (b) 37°C. Method: (a, b) Static-desiccator (saturated salt solutions); (c) quartz spring balance sorption apparatus. (Karel and Nickerson, 1964.)

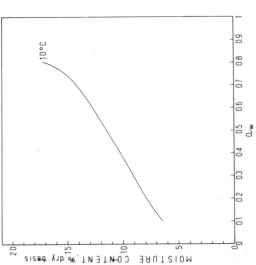

Fig. 279. Onion seed (sorption, 10°C): Morley 1938 crop from Burt Leonard, West Row; initial water content 11.5% (dry basis). Method: Static-jar (sulfuric acid solutions). (Gane, 1948.)

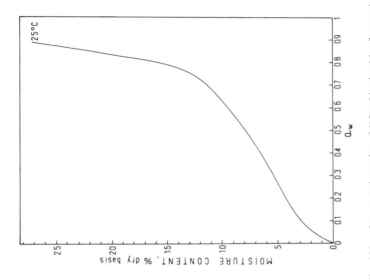

Fig. 282. Orgeat (adsorption, 25°C): Obtained by freeze-drying a 20°C Brix extract. **Method:** Static-desiccator (saturated salt solutions). (Piñaga and Lafuente, 1965.)

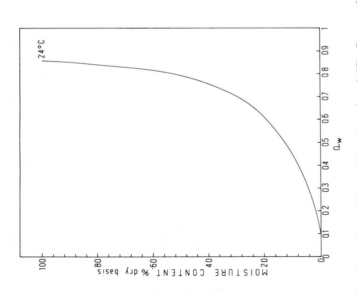

Fig. 281. Orange juice solids (adsorption, 24°C): Commercial dehydrated (spray-dried) orange juice. **Method:** Electrobalance assembly. (Berlin et al., 1973b.)

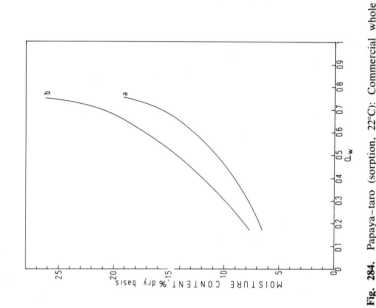

Fig. 283. Ovalbumin (adsorption, 21°C): Ovalbumin was lipid free, three times recrystalized, and freeze-dried (Worthington Bio-chemicals). Method: Static-desiccator (sulfuric acid/saturated salt solutions). (Hansen, 1976.)

Fig. 284. Papaya–taro (sorption, 22°C): Commercial whole white taro (Apii variety) corms and papaya puree were combined for drum drying; papaya–taro ratio: (a) 1:2, (b) 3:2. Method: Static (saturated salt solutions). (Nip, 1978.)

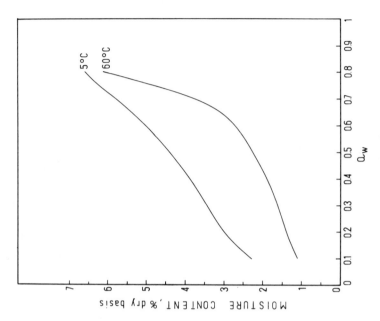

Fig. 285. Paranut (adsorption): Samples were vacuum-dried at 30°C before adsorption; (a) 5, (b) 25, (c) 60°C. Method: Static-jar (sulfuric acid solutions). (Wolf *et al.*, 1973.)

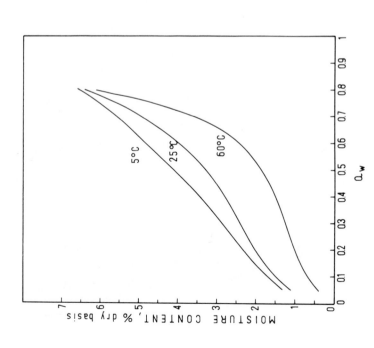

Fig. 286. Paranut (desorption): Samples were vacuum-dried at 30°C before adsorption; desorption following humidification at 90% relative humidity. Method: Static-jar (sulfuric acid solutions). (Wolf *et al.*, 1973.)

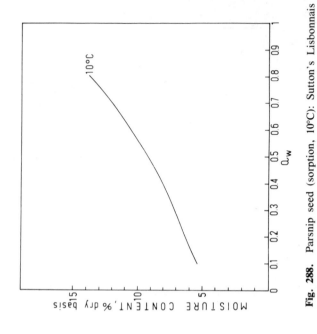

Fig. 288. Parsnip seed (sorption, 10°C): Sutton's Lisbonnais 1938 crop, initial moisture content 10.2% (dry basis). Method: Static-jar (sulfuric acid solutions). (Gane, 1948.)

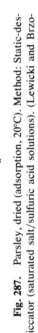

Fig. 287. Parsley, dried (adsorption, 20°C). Method: Static-desiccator (saturated salt/sulfuric acid solutions). (Lewicki and Brzozowski, 1973.)

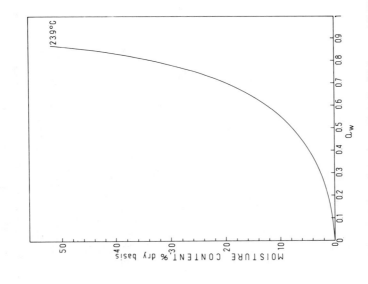

Fig. 290. Peach (adsorption, 23.9°C). [Schwarz (1943), as quoted by Henderson (1952).]

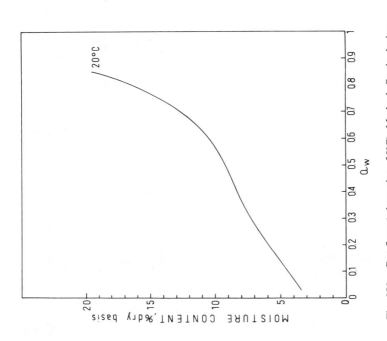

Fig. 289. Pea flour (adsorption, 20°C). Method: Static-desiccator (saturated salt/sulfuric acid solutions). (Lewicki and Brzozowski, 1973.)

155

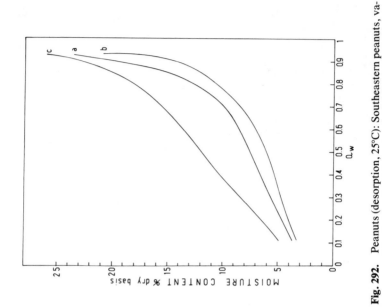

Fig. 292. Peanuts (desorption, 25°C): Southeastern peanuts, variety Runner; (a) whole pod, (b) kernel, (c) shell. [Karon and Hillery (1949), as quoted by Agrawal *et al.*, (1969).]

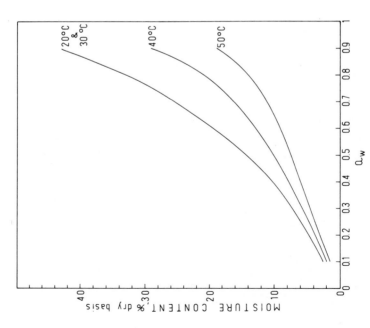

Fig. 291. Peach (adsorption): Freeze-dried disks 1 mm thick were used; (a) 20, 30, (b) 40, (c) 50°C. Method: Vacuum sorption apparatus with quartz spring balance. (Saravacos and Stinchfield, 1965.)

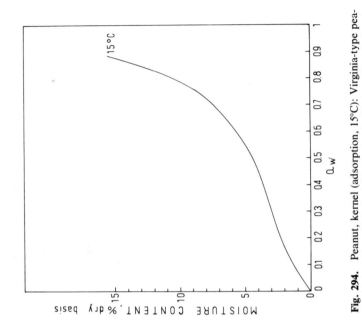

Fig. 294. Peanut, kernel (adsorption, 15°C): Virginia-type peanuts, NC-5 variety. Method: Static-jar (sulfuric acid solutions). (Young, 1976.)

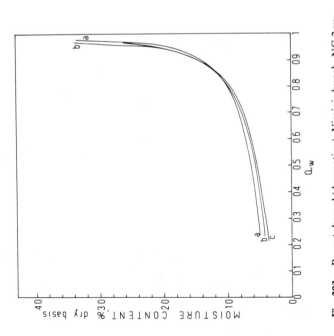

Fig. 293. Peanut, kernel (desorption): Virginia bunch, NC-2 variety; (a) 10, (b) 21.1, (c) 32.2°C. [Beasley (1962), as quoted by Agrawal et al. (1969).]

157

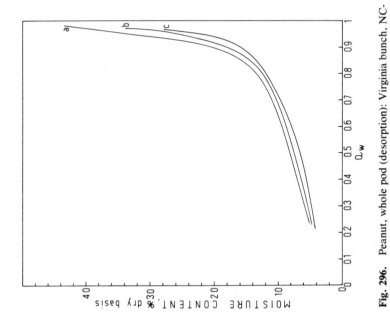

Fig. 296. Peanut, whole pod (desorption): Virginia bunch, NC-2 variety; (a) 10, (b) 21.1, (c) 32.2°C. [Beasley (1962), as quoted by Agrawal et al., 1969).]

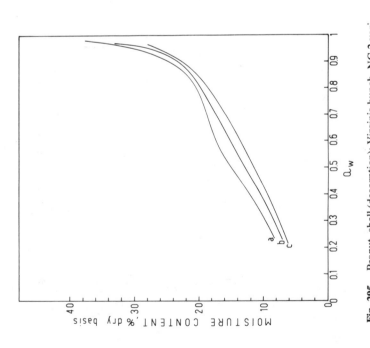

Fig. 295. Peanut, shell (desorption): Virginia bunch, NC-2 variety; (a) 10, (b) 21.1, (c) 32.2°C. [Beasley (1962), as quoted by Agrawal et al., (1969).]

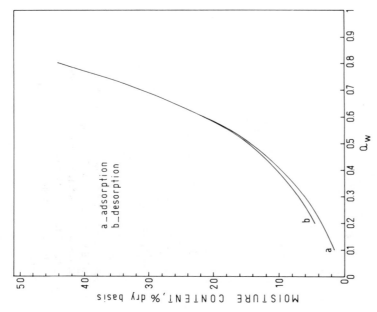

Fig. 298. Pear (adsorption and desorption, 25°C): Air-dried and vacuum-dried at 30°C before adsorption; desorption following humidification at 90% relative humidity. Method: Static-jar (sulfuric acid solutions). (Wolf *et al.*, 1973.)

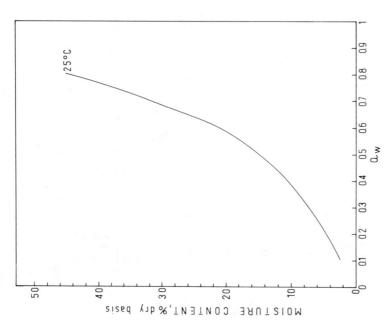

Fig. 297. Pear (adsorption and desorption, 25°C): freeze-dried and vacuum-dried at 30°C before adsorption; desorption following humidification at 90% relative humidity. Method: Static-jar (sulfuric acid solutions). (Wolf *et al.*, 1973.)

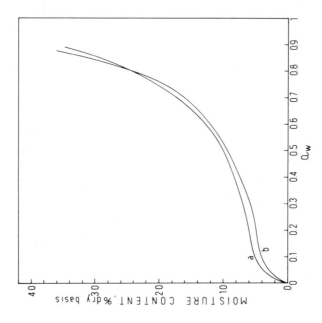

Fig. 300. Peas [(a) desorption, 19.5°C; (b) adsorption, 25°C]: (a) A freeze-dried sample was allowed to equilibrate to about 20% (dry basis) moisture content before isotherm determination; (b) freeze-dried samples. Method: (a) Manometric apparatus; (b) static-desiccator (saturated salt solutions). [(a) Taylor (1961); (b) Lafuente and Piñaga (1966).]

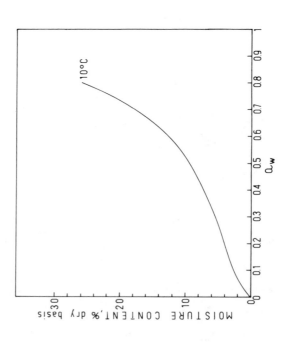

Fig. 299. Peas (adsorption, 10°C): Scalded and freeze-dried before adsorption. Method: Static-jar (sulfuric acid solutions). (Gane, 1950.)

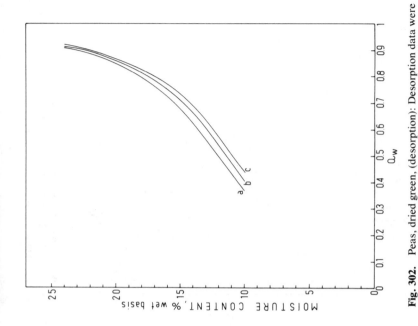

Fig. 301. Peas, dried green (adsorption): Samples were air-dried at 30°C to bring a_w below 0.30 before isotherm determination; (a) 15, (b) 25, (c) 35°C. Method: Dew-point. (Pixton and Henderson, 1979.)

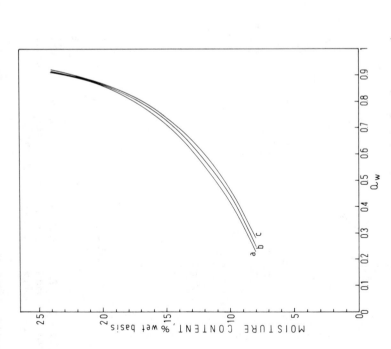

Fig. 302. Peas, dried green, (desorption): Desorption data were obtained after raising a_w to 0.95 by adding liquid water; (a) 15, (b) 25, (c) 35°C. Method: Dew-point. (Pixton and Henderson, 1979.)

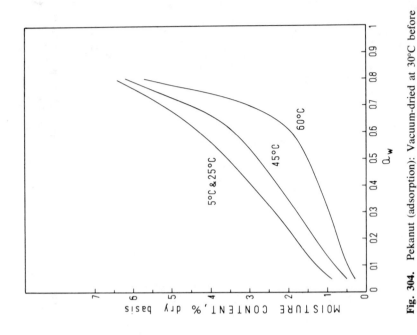

Fig. 304. Pekanut (adsorption): Vacuum-dried at 30°C before adsorption; (a) 5, 25, (b) 45, (c) 60°C. Method: Jar with air agitation (sulfuric acid solutions). (Wolf *et al.*, 1973.)

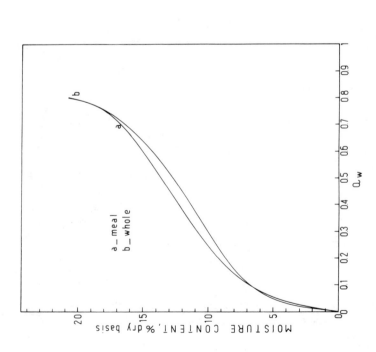

Fig. 303. Pea, seeds (sorption, 10°C): Sutton's Improved Pilot 1938 crop; initial moisture content 13.3% (dry basis). Method: Static-jar (sulfuric acid solutions). (Gane, 1948.)

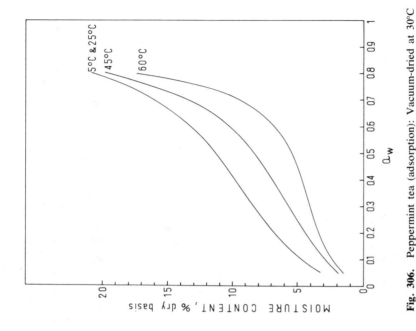

Fig. 306. Peppermint tea (adsorption): Vacuum-dried at 30°C before adsorption; (a) 5, 25, (b) 45, (c) 60°C. Method: Jar with air agitation (sulfuric acid solutions). (Wolf *et al.*, 1973.)

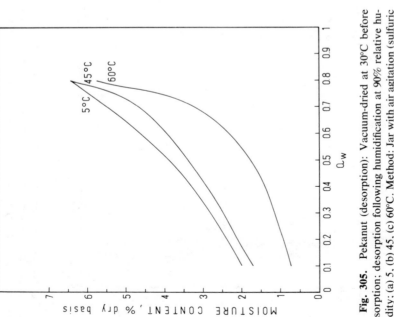

Fig. 305. Pekanut (desorption): Vacuum-dried at 30°C before adsorption; desorption following humidification at 90% relative humidity; (a) 5, (b) 45, (c) 60°C. Method: Jar with air agitation (sulfuric acid solutions). (Wolf *et al.*, 1973.)

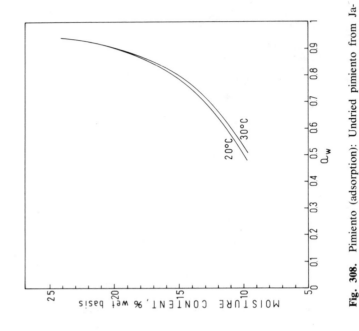

Fig. 308. Pimiento (adsorption): Undried pimiento from Jamaica, dried to about 8% for determining the adsorption isotherm. Method: Dew-point. (Ayerst, 1965.)

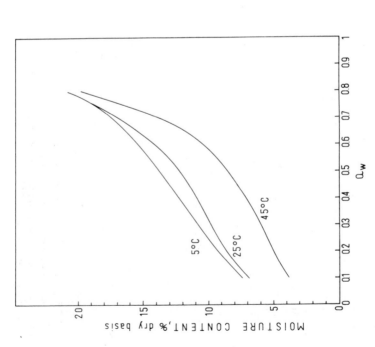

Fig. 307. Peppermint tea (desorption): Vacuum-dried at 30°C before adsorption; desorption following humidification at 90% relative humidity; (a) 5, (b) 25, (c) 45°C. Method: Jar with air agitation (sulfuric acid solutions). (Wolf et al., 1973.)

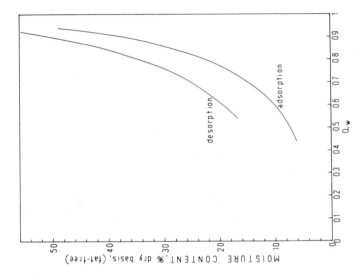

Fig. 310. Pistacho nut, hull (adsorption and desorption, 20°C): Nuts from Turkey with moisture content 5–7% (wet basis); hulls contained a mean lipid content of 8.6% (wet basis) and were allowed to equilibrate with water to a moisture content of 50–60% before desorption. Method: Manometric apparatus. (Denizel *et al.*, 1976.)

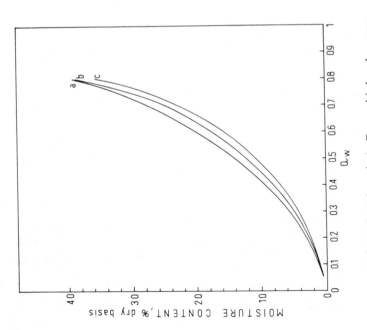

Fig. 309. Pineapple (adsorption): Freeze-dried and vacuum-dried at 30°C before adsorption; (a) 5, 25, (b) 45, (c) 60°C. Method: Jar with air agitation (sulfuric acid solutions). (Wolf *et al.*, 1973.)

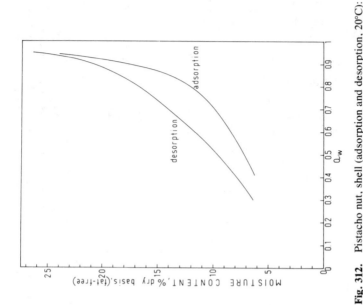

Fig. 311. Pistacho nut, kernel (adsorption and desorption, 20°C): Nuts from Turkey with moisture content 5–7% (wet basis); kernels contained a mean lipid content of 56.5% (wet basis) and were allowed to equilibrate with water to a moisture content of 25–30% (wet basis) before desorption. Method: Manometric apparatus. (Denizel *et al.*, 1976.)

Fig. 312. Pistacho nut, shell (adsorption and desorption, 20°C): Nuts from Turkey with moisture content 5–7% (wet basis); shells were allowed to equilibrate with water to a moisture content of 25–30% (wet basis) before desorption. Method: Manometric apparatus. (Denizel *et al.*, 1976.)

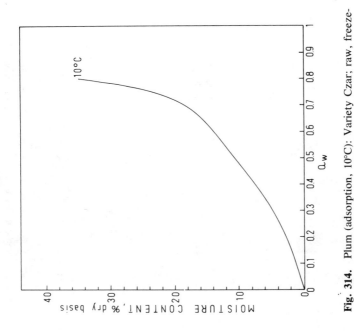

Fig. 314. Plum (adsorption, 10°C): Variety Czar: raw, freeze-dried. Method: Static-jar (sulfuric acid solutions). (Gane, 1950.)

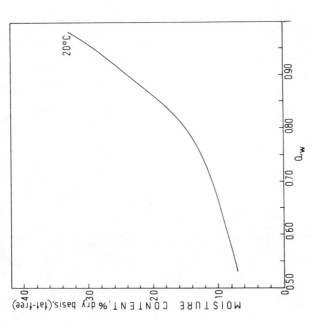

Fig. 313. Pistacho nuts, whole (adsorption, 20°C): Nuts from Turkey. Method: Manometric apparatus. (Denizel *et al.*, 1976.)

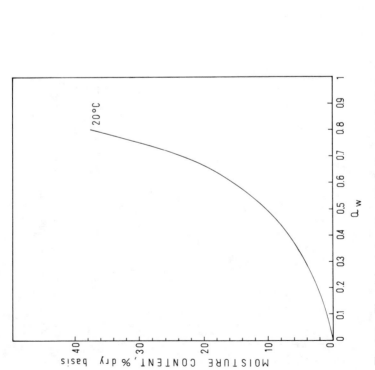

Fig. 316. Pork, cooked (adsorption and desorption, 4.4°C): Freeze-dried. Method: Vacuum sorption apparatus with quartz spring balance. (Wolf *et al.*, 1972.)

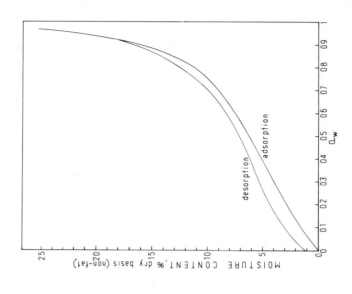

Fig. 315. Plum juice (adsorption, 20°C): Spray-dried. Method: Static-desiccator (saturated salt/sulfuric acid solutions). (Lewicki, 1976.)

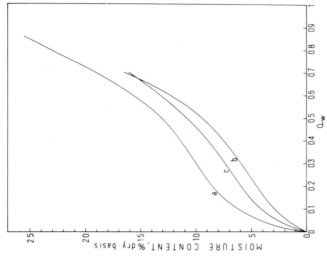

Fig. 318. Potato (adsorption and desorption): (a) Freeze-dried sample, ground and allowed to equilibrate to about 20% (dry basis) moisture content before isotherm determination: desorption, 19.5°C; (b) desorption and adsorption, 25°C; (c) variety Russet; dried in vacuum oven at 70°C before adsorption; desorption and adsorption, 37°C. Method: (a) Manometric apparatus; (b) static-desiccator (saturated salt solutions); (c) static-desiccator (sulfuric acid solutions). [(a) Taylor (1961); (b) Iglesias (1975); (c) Makower and Dehority (1943).]

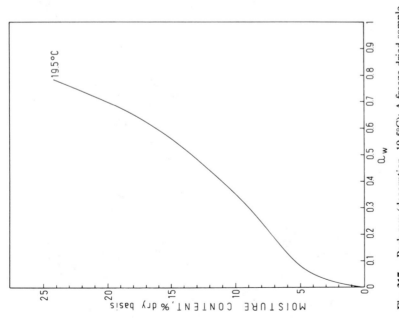

Fig. 317. Pork raw (desorption, 19.5°C): A freeze-dried sample was ground and allowed to equilibrate at about 20% (dry basis) before isotherm determination. Method: Manometric apparatus. (Taylor, 1961.)

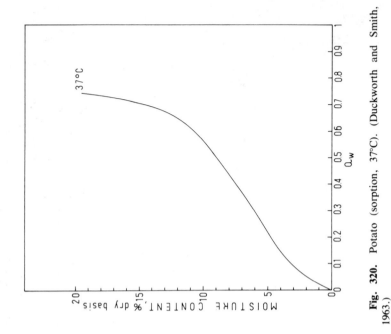

Fig. 320. Potato (sorption, 37°C). (Duckworth and Smith, 1963.)

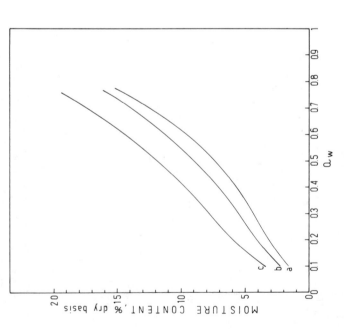

Fig. 319. Potato (adsorption, 30°C): (a) Samples air-dried at 71.7°C; (b) samples puff-dried; (c) samples freeze-dried at room temperature. Method: Vacuum sorption apparatus with quartz spring balance. (Saravacos, 1967.)

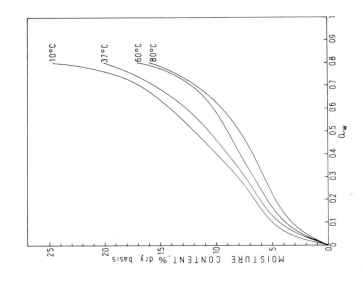

Fig. 322. Potato (sorption): Scalded and air-dried: initial moisture content 5.2% (dry basis); (a) 10, (b) 37, (c) 60, (d) 80°C. Method: Static-jar (sulfuric acid solutions). (Gane, 1950.)

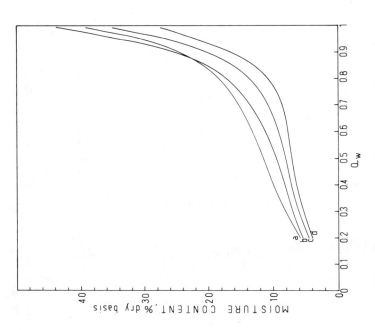

Fig. 321. Potato (adsorption): Disks 1 mm thick were cut from blanched potato and freeze-dried at 30°C: (a) 20, (b) 30, (c) 40, (d) 50°C. Method: Vacuum sorption apparatus with quartz spring balance. (Saravacos and Stinchfield, 1965.)

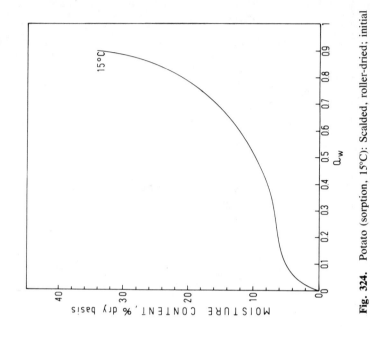

Fig. 324. Potato (sorption, 15°C): Scalded, roller-dried; initial moisture content 7.9% (dry basis). Method: Static-jar (sulfuric acid solutions). (Gane, 1950.)

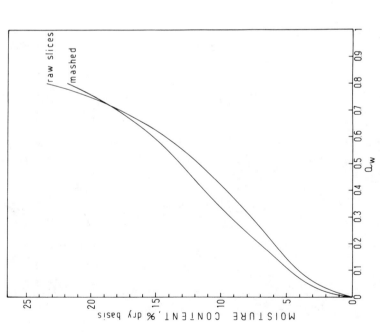

Fig. 323. Potato (adsorption, 10°C): (a) Raw slices, imported from Jersey (immature); (b) mashed, variety King Edward. Method: Static-jar (sulfuric acid solutions). (Gane, 1950.)

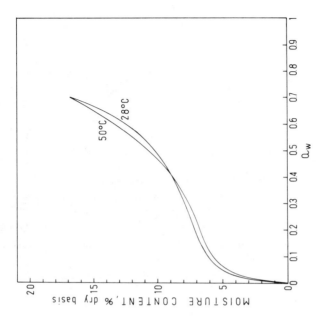

Fig. 325. Potato, cooked (sorption): Cooked, air-dried; initial moisture content 7.2% (dry basis). Method: Static-jar (sulfuric acid solutions). (Gane, 1950.)

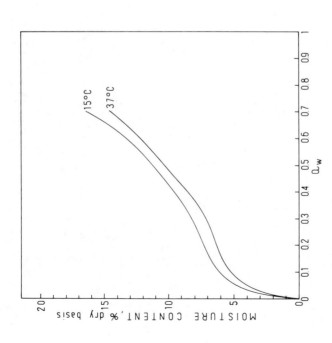

Fig. 326. Potato, Canadian (sorption): Scalded, air-dried; initial moisture content 8.6% (dry basis); sugar content 9.7% (dry basis). Method: Static-jar (sulfuric acid solutions). (Gane, 1950.)

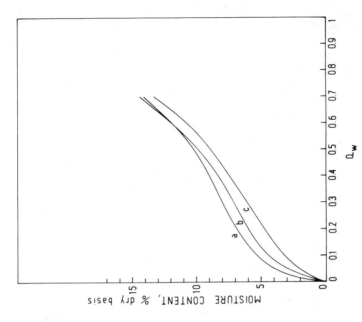

Fig. 328. Potato, English (sorption): Scalded, air-dried, initial moisture content 7.7% (dry basis); sugar content 3.7% (dry basis); (a) 28, (b) 50, (c) 70°C. Method: Static-jar (sulfuric acid solutions). (Gane, 1950.)

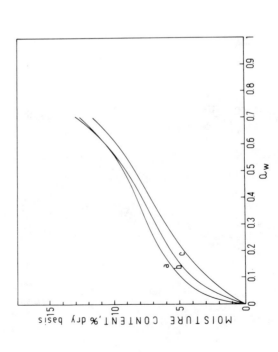

Fig. 327. Potato, English (sorption): Scalded, air-dried, initial moisture content 8.2% (dry basis); sugar content 0.7% (dry basis); (a) 28, (b) 50, (c) 70°C. Method: Static-jar (sulfuric acid solutions). (Gane, 1950.)

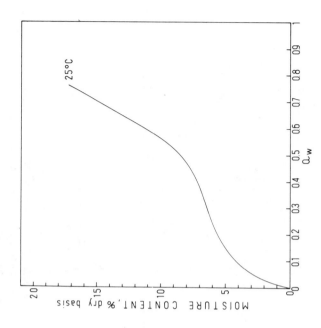

Fig. 330. Potato Flakes (adsorption, 25°C): Data are average values for following varieties: Katahdin (New York, Pennsylvania), White Rose (California), Pontiae (Minnesota), Russet (Maine, Idaho, Washington), and Cobbler (Minnesota). Method: Static-desiccator (saturated salt solutions). (Strolle and Cording, 1965.)

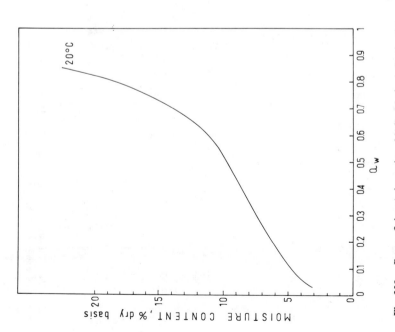

Fig. 329. Potato flakes (adsorption, 20°C). Method: Static-disiccator (saturated salt/sulfuric acid solutions). (Lewicki and Brzozowski, 1973.)

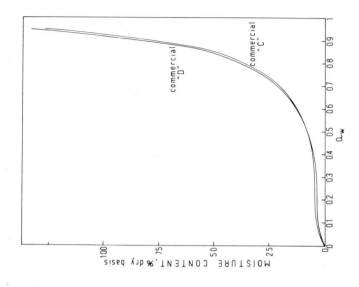

Fig. 332. Prunes (sorption, 10°C): (a) Commercial "C": initial moisture content 26.0% (dry basis); (b) commercial "D": initial moisture content 43.2% (dry basis). Method: Static-jar (sulfuric acid solutions). (Gane, 1950.)

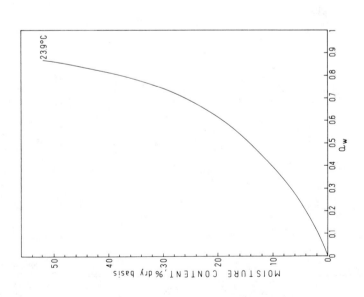

Fig. 331. Prunes (adsorption, 23.9°C). [Schwarz (1943), as quoted by Henderson (1952).]

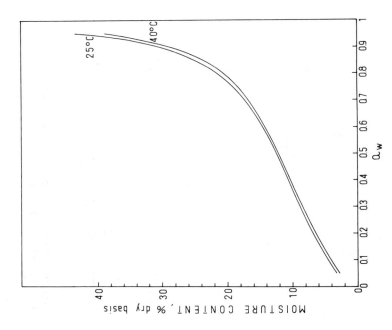

Fig. 334. γ-Pseudoglobulin (adsorption). Method: Static-desiccator (sulfuric acid solutions); portion of serum precipitated by 34% saturated ammonium sulfate; proteins dialyzed against pure water and air-dried. (Bull, 1944.)

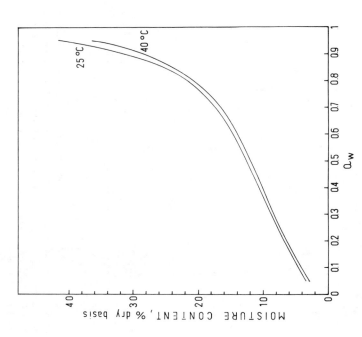

Fig. 333. α- and β-Pseudoglobulin (adsorption): Method: Static-desiccator (sulfuric acid solutions); portion of serum precipitated between 34 and 45% by saturated ammonium sulfate; proteins dialyzed against pure water and air-dried. (Bull, 1944.)

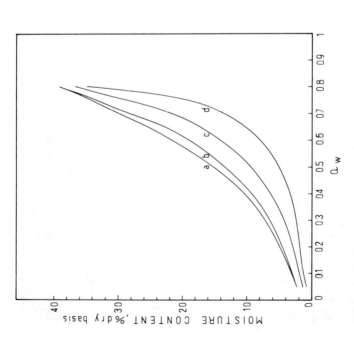

Fig. 335. Radish (absorption): Freeze-dried and vacuum-dried at 30°C before adsorption; adsorption isotherm at 25°C is coincident with desorption at same temperature: (a) 5, (b) 25, (c) 45, (d) 60°C. Method: Jar with air agitation (sulfuric acid solutions). (Wolf *et al.*, 1973.)

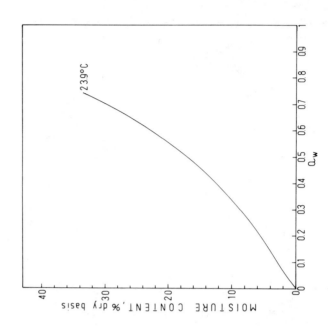

Fig. 336. Raisins (adsorption, 23.9°C). [Schwarz (1943), as quoted by Henderson (1952).]

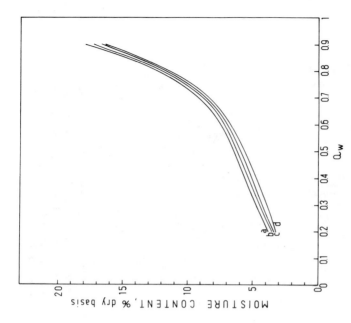

Fig. 337. Rapeseed, Gulle (adsorption): European variety Gulle, spring-sown in Great Britain, harvested in 1973; samples were air-dried at 30°C before adsorption; (a) 5, (b) 15, (c) 25, (d) 35°C. Method: Dew-point. (Pixton and Warburton, 1977a.)

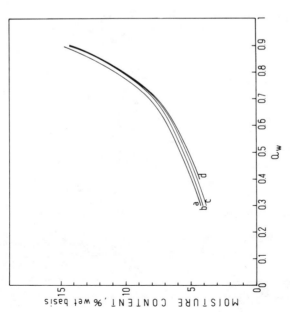

Fig. 338. Rapeseed, Gulle (desorption): European variety Gulle, spring-sown in Great Britain, harvested in 1973; to obtain desorption curve, the moisture content of previously dried rapeseed was raised by adding water; (a) 5, (b) 15, (c) 25, (d) 35°C. Method: Dew-point. (Pixton and Warburton, 1977a.)

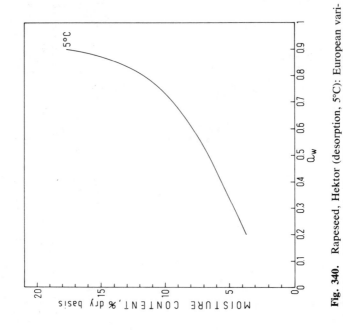

Fig. 340. Rapeseed, Hektor (desorption, 5°C): European variety Hektor, winter-sown in Great Britain, harvested in 1973; to obtain desorption curve, the moisture content of previously dried rapeseed was raised by adding water. Method: Dew-point. (Pixton and Warburton, 1977a.)

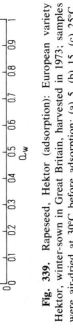

Fig. 339. Rapeseed, Hektor (adsorption): European variety Hektor, winter-sown in Great Britain, harvested in 1973; samples were air-dried at 30°C before adsorption: (a) 5, (b) 15, (c) 25°C. Method: Dew-point. (Pixton and Warburton, 1977a.)

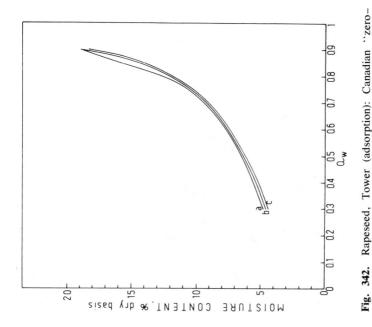

Fig. 342. Rapeseed, Tower (adsorption): Canadian "zero-zero" rapeseed variety Tower, harvested in 1974; samples were air-dried at 30°C before adsorption; (a) 5, (b) 15, (c) 25°C. Method: Dew-point. (Pixton and Warburton, 1977a.)

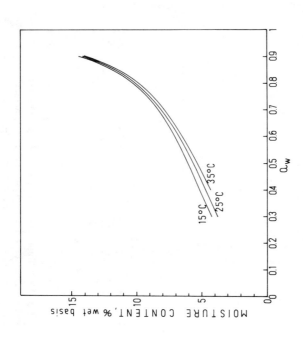

Fig. 341. Rapeseed, Hektor (desorption): European variety Hektor, winter-sown in Great Britain, harvested in 1973; to obtain desorption curves, the moisture content of previously dried rapeseed was raised by adding water; (a) 15, (b) 25, (c) 35°C. Method: Dew-point. (Pixton and Warburton, 1977a.)

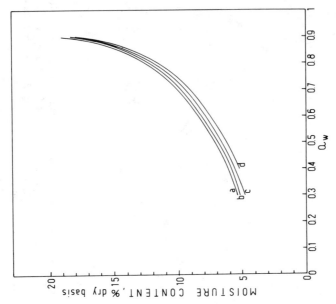

Fig. 343. Rapeseed (adsorption, 35°C): European variety Hektor (b), winter-sown in Great Britain, harvested in 1973, and Canadian "zero–zero" rapeseed variety Tower (a), harvested in 1974; samples were air-dried at 30°C before adsorption. Method: Dewpoint. (Pixton and Warburton, 1977a.)

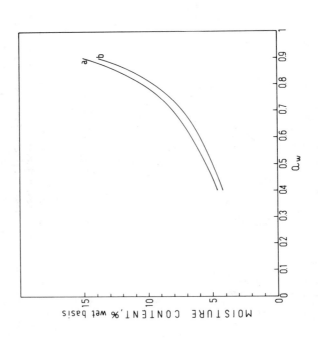

Fig. 344. Rapeseed, Tower (desorption): Canadian "zero–zero" rapeseed variety Tower, harvested in 1974; to obtain desorption curves, the moisture content of previously dried rapeseed was raised by adding water; (a) 5, (b) 15, (c) 25, (d) 35°C. Method: Dewpoint. (Pixton and Warburton, 1977a.)

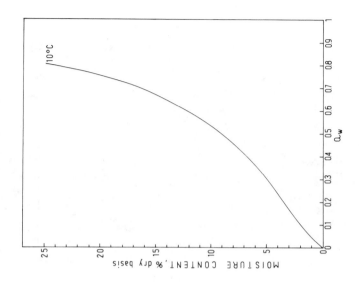

Fig. 346. Rhubarb (adsorption, 10°C). Method: Static-jar (sulfuric acid solutions). (Gane, 1950.)

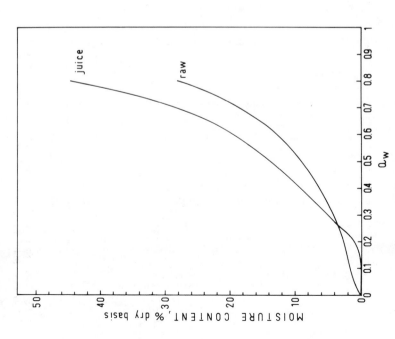

Fig. 345. Raspberry (adsorption, 10°C): Soluble constituents 45% (dry basis); (a) juice, (b) raw. Method: Static-jar (sulfuric acid solutions). (Gane, 1950.)

183

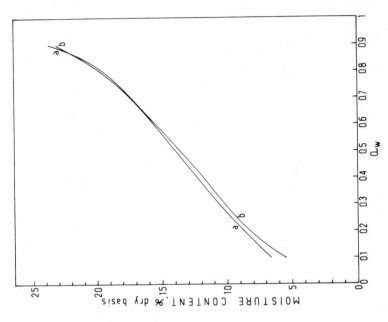

Fig. 348. Rice (25°C): (a) Brown, (b) white. [(a) Wink and Sears (1950), (b) Coleman and Fellows (1925), as quoted by Ferrel *et al.* (1966).]

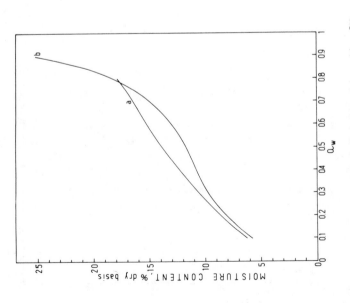

Fig. 347. Rice (25°C): (a) Parboiled, (b) quick cooking. [Karon and Adams (1949), as quoted by Ferrel *et al.*, (1966).]

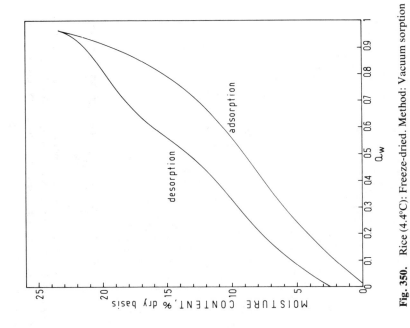

Fig. 349. Rice, cooked (desorption, 19.5°C): A freeze-dried sample was ground and allowed to equilibrate at about 20% (dry basis) moisture content before isotherm determination. Method: Manometric apparatus. (Taylor, 1961.)

Fig. 350. Rice (4.4°C): Freeze-dried. Method: Vacuum sorption apparatus with quartz spring balance. (Wolf *et al.*, 1972.)

185

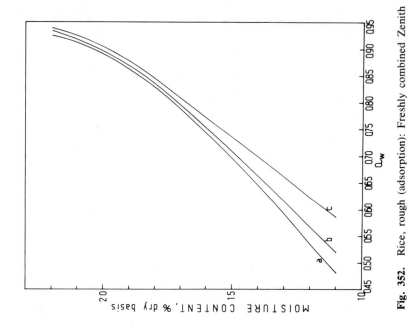

Fig. 352. Rice, rough (adsorption): Freshly combined Zenith rice from Crowley, LA: (a) 26.7; (b) 34.4, (c) 43.9°C. Method: Dynamic (chamber with circulation of air through the sample). (Hogan and Karon, 1955.)

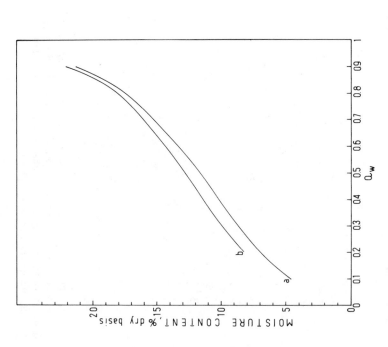

Fig. 351. Rice, rough (25°C). [(a) Coleman and Fellows (1925), as quoted by Ferrel *et al.* (1966); (b) Wink and Sears (1950), as quoted by Ferrel *et al.* (1966).]

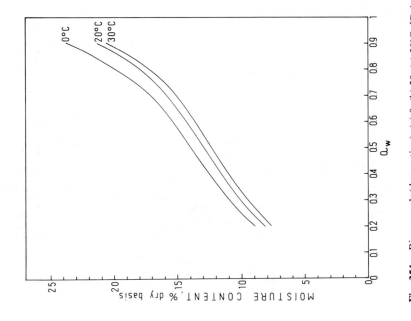

Fig. 354. Rice, rough (desorption): (a) 0, (b) 20, (c) 30°C. [Bakharev (1960), as quoted by Agrawal *et al.* (1969).]

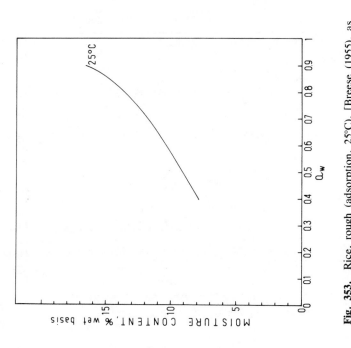

Fig. 353. Rice, rough (adsorption, 25°C). [Breese (1955), as quoted by Juliano (1964).]

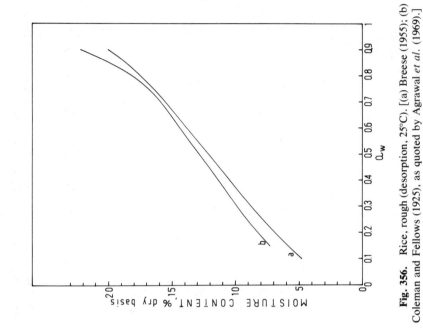

Fig. 356. Rice, rough (desorption, 25°C). [(a) Breese (1955); (b) Coleman and Fellows (1925), as quoted by Agrawal et al. (1969).]

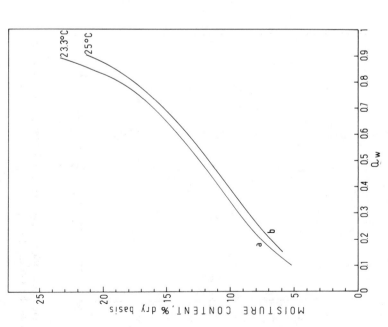

Fig. 355. Rice, rough (desorption): (a) Colusa rice, 23.3°C; (b) 25°C. [(a) Henderson (1969); (b) Karon and Adams (1949), as quoted by Agrawal et al. (1969).]

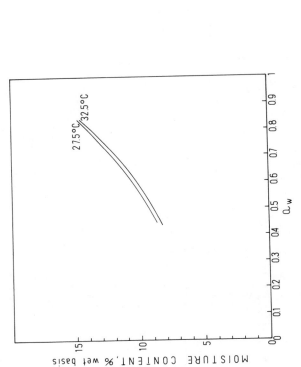

Fig. 358. Rice, rough (desorption): Nonwaxy variety Peta (indica); samples equilibrated to $a_w = 0.965$ before desorption determinations. Method: Dynamic (chamber with circulation of air over the sample). (Juliano, 1964.)

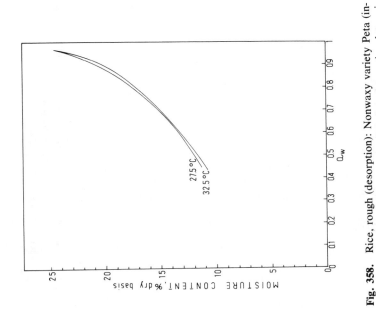

Fig. 357. Rice, rough (adsorption): Nonwaxy variety Peta (indica); samples dried at room temperature prior to adsorption. Method: Dynamic (chamber with circulation of air over the sample). (Juliano, 1964.)

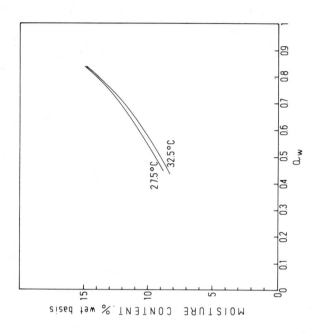

Fig. 360. Rice, rough (adsorption): Nonwaxy variety Taichung 65 (japonica); samples dried at room temperature before adsorption. Method: Dynamic (chamber with circulation of air over the sample). (Juliano, 1964.)

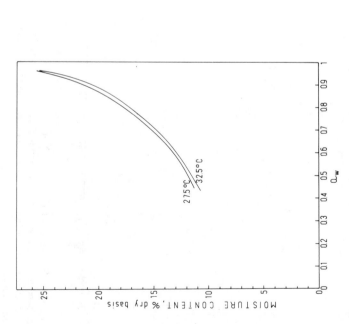

Fig. 359. Rice rough, (desorption): Nonwaxy variety Taichung 65 (japonica); samples equilibrated to $a_w = 0.965$ before desorption determinations. Method: Dynamic (chamber with circulation of air over the sample). (Juliano, 1964.)

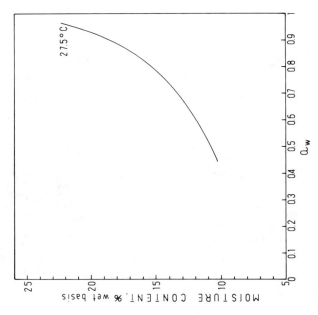

Fig. 361. Rice, rough (adsorption): Waxy variety Malagkit Sungsong (indica); samples dried at room temperature before adsorption. Method: Dynamic (chamber with circulation of air over the sample). (Juliano, 1964.)

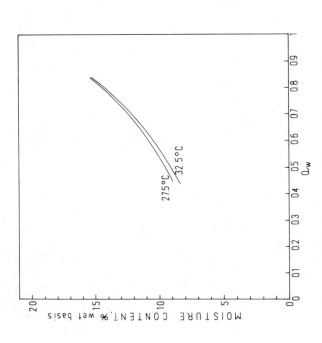

Fig. 362. Rice, rough (desorption, 27.5°C): Waxy variety Malagkit Sungsong (indica); samples equilibrated to $a_w = 0.965$ before desorption determinations. Method: Dynamic (chamber with circulation of air over the sample). (Juliano, 1964.)

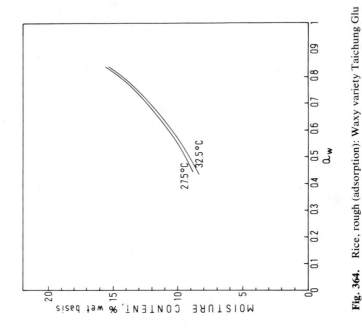

Fig. 364. Rice, rough (adsorption): Waxy variety Taichung Glu 46 (japonica); samples dried at room temperature before adsorption. Method: Dynamic (chamber with circulation of air over the sample). (Juliano, 1964.)

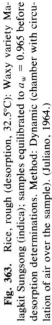

Fig. 363. Rice, rough (desorption, 32.5°C): Waxy variety Malagkit Sungsong (indica); samples equilibrated to $a_w = 0.965$ before desorption determinations. Method: Dynamic (chamber with circulation of air over the sample). (Juliano, 1964.)

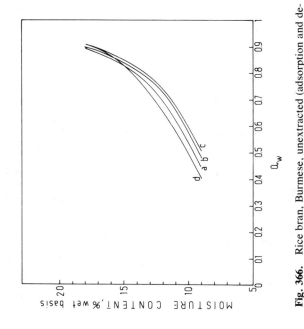

Fig. 366. Rice bran, Burmese, unextracted (adsorption and desorption): Oil content 11% (dry basis); samples dried at room temperature prior to adsorption; for desorption, moisture content was previously raised by adding liquid water; adsorption: (a) 15, (b) 25, (c) 35°C; desorption: (d) 35°C. Method: Dew-point. (Pixton and Warburton, 1975b.)

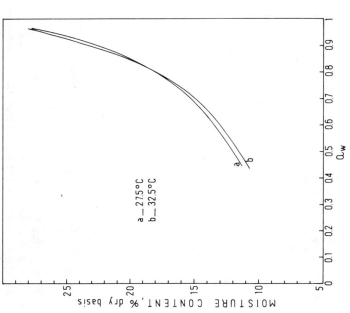

Fig. 365. Rice, rough (desorption): Waxy variety Taichung Glu 46 (japonica); samples equilibrated to $a_w = 0.965$ before desorption determinations. Method: Dynamic (chamber with circulation of air over the sample). (Juliano, 1964.)

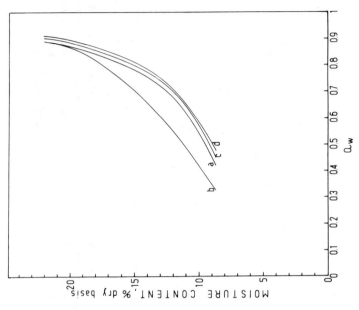

Fig. 367. Rice bran, Burmese, unextracted (desorption): Oil content 11% (dry basis); for desorption, moisture content of samples was previously raised by adding liquid water. Method: Dew-point. (Pixton and Warburton, 1975b.)

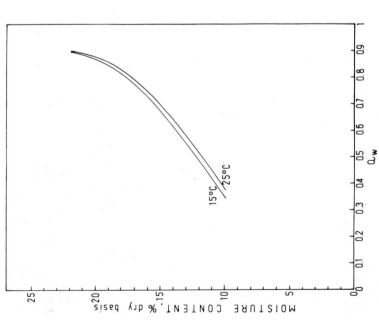

Fig. 368. Rice bran, Indian, extracted (adsorption and desorption): Oil content 0.6% (dry basis); samples dried at room temperature prior to adsorption; for desorption, moisture content was previously raised by adding liquid water; adsorption: (a) 15°C, (c) 25°C, (d) 35°C; desorption: (b) 15°C. Method: Dew-point. (Pixton and Warburton, 1975b.)

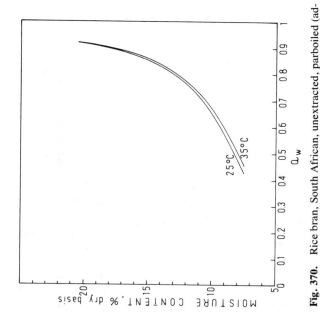

Fig. 369. Rice bran, Indian, extracted (desorption): Oil content 0.6% (dry basis); prior to desorption moisture content was raised by adding liquid water. Method: Dew-point. (Pixton and Warburton, 1975b.)

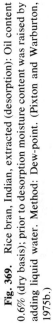

Fig. 370. Rice bran, South African, unextracted, parboiled (adsorption): Oil content 29.7% (dry basis); samples dried at room temperature prior to adsorption. Method: Dew-point. (Pixton and Warburton, 1975b.)

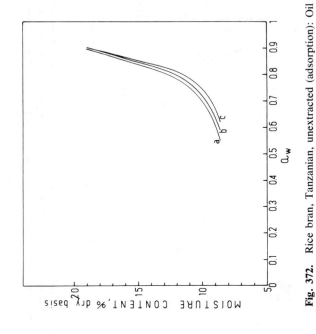

Fig. 371. Rice bran, South African, unextracted, parboiled (adsorption and desorption): Oil content 19.7% (dry basis): samples dried at room temperature prior to adsorption; for desorption, moisture content was previously raised by adding liquid water: adsorption: (a) 15°C; desorption: (b) 15, (c) 25, (d) 35°C. Method: Dew-point. (Pixton and Warburton, 1975b)

Fig. 372. Rice bran, Tanzanian, unextracted (adsorption): Oil content 19.5% (dry basis); samples dried at room temperature prior to adsorption; (a) 15, (b) 25, (c) 35°C. Method: Dew-point. (Pixton and Warburton, 1975b.)

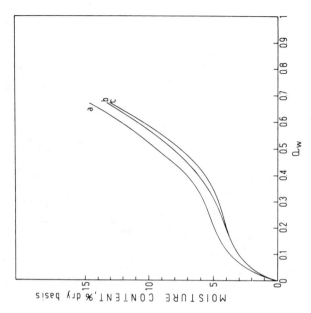

Fig. 374. *Sacc. cerevisiae* (adsorption and desorption, 20°C): *Sacc. cerevisiae* strain ATCC 9763-1 N; spray-dried with 170°C inlet air temperature; for desorption a cell suspension was used; (a) adsorption, spray-dried; (b) adsorption, freeze-dried; (c) desorption. Method: Static-desiccator (saturated salt solutions). (Peri and De Cesari, 1974.)

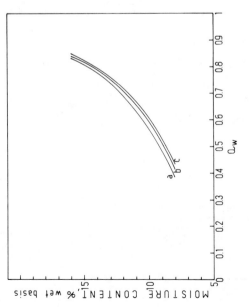

Fig. 373. Rice bran, Tanzanian, unextracted (desorption): Oil content 19.5% (dry basis); prior to desorption moisture content of samples was raised by adding liquid water; (a) 15, (b) 25, (c) 35°C. Method: Dew-point. (Pixton and Warburton, 1975b.)

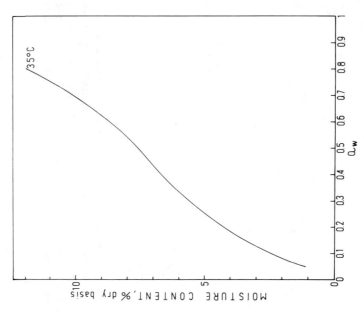

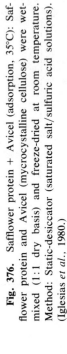

Fig. 376. Safflower protein + Avicel (adsorption, 35°C): Safflower protein and Avicel (mycrocystalline cellulose) were wet-mixed (1:1 dry basis) and freeze-dried at room temperature. Method: Static-desiccator (saturated salt/sulfuric acid solutions). (Iglesias *et al.*, 1980.)

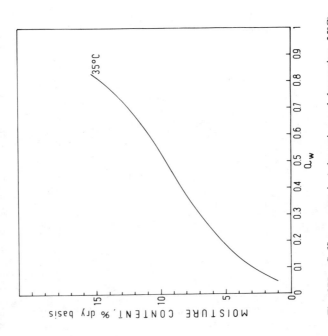

Fig. 375. Safflower protein (adsorption and desorption, 35°C): Isolated safflower protein sample freeze-dried at room temperature. Method: Static-desiccator (saturated salt/sulfuric acid solutions). (Iglesias *et al.*, 1980.)

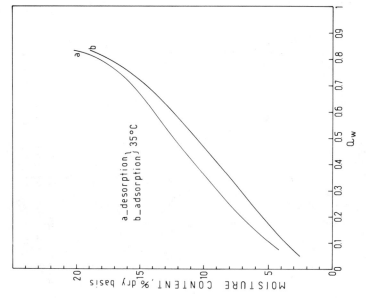

Fig. 378. Safflower protein + starch gel (adsorption and desorption, 35°C): Isolated safflower protein and starch were wet-mixed (1:1 dry basis) with heating and freeze-dried at room temperature. Method: Static-desiccator (saturated salt/sulfuric acid solutions). (Iglesias *et al.*, 1980.)

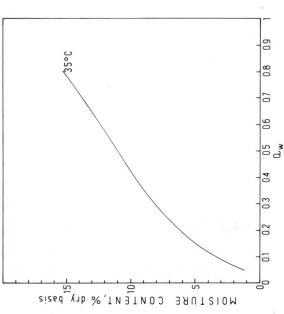

Fig. 377. Safflower protein + starch (adsorption, 35°C): Isolated safflower protein and starch were wet-mixed (1:1 dry basis) and freeze-dried at room temperature. Method: Static-desiccator (saturated salt/sulfuric acid solutions). (Iglesias *et al.*, 1980.)

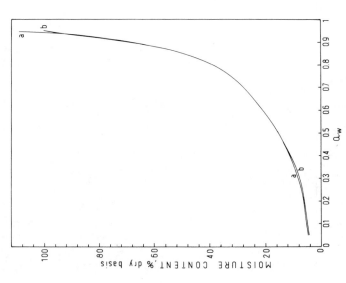

Fig. 380. Salmon (adsorption, 37°C): Pacific sockeye salmon (*Onocorhynchus nerka*) freeze-dried at room temperature. Method: Static-desiccator (saturated salt solutions). (Martinez and Labuza, 1968.)

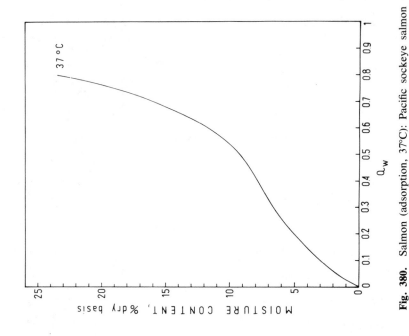

Fig. 379. Salmin (adsorption): Salmin sulfate was electrodialyzed, freeze-dried, and used as the base; (a) 25, (b) 40°C. Method: Static-desiccator (sulfuric acid solutions). (Bull, 1944.)

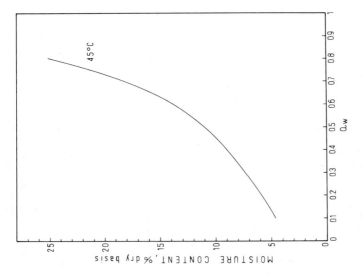

Fig. 382. Salsify (desorption, 45°C): Freeze-dried and vacuum-dried at 30°C before adsorption; desorption following humidification at 90% relative humidity. **Method:** Jar with air agitation (sulfuric acid solutions). (Wolf *et al.*, 1973.)

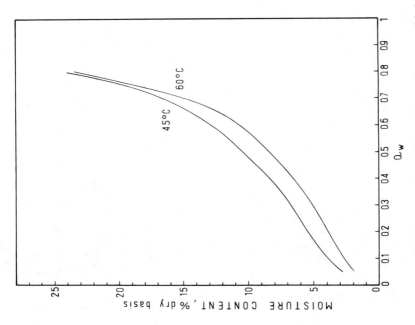

Fig. 381. Salsify (adsorption): Freeze-dried and vacuum-dried at 30°C before adsorption. **Method:** Jar with air agitation (sulfuric acid solutions). (Wolf *et al.*, 1973.)

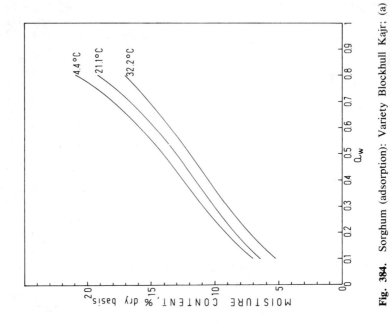

Fig. 383. Sarcoplasmic fraction, beef (adsorption and desorption): Obtained from the longissimus dorsi muscle of beef; desorption isotherm was obtained after humidification at about 99.5% relative humidity; adsorption: (c) 21.1°C; desorption: (a) 21.1, (b) 11.1°C. Method: Electrobalance assembly. (Palnitkar and Heldman, 1971.)

Fig. 384. Sorghum (adsorption): Variety Blockhull Kajr; (a) 4.4, (b) 21.1, (c) 32.2°C. [Fenton (1941), as quoted by Chen (1971).]

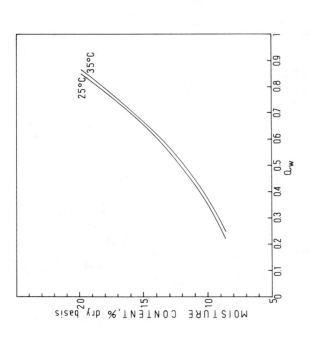

Fig. 386. Sorghum (desorption): Variety Bukura Mahemba from Tanganyika. Method: Dew-point. (Ayerst, 1965.)

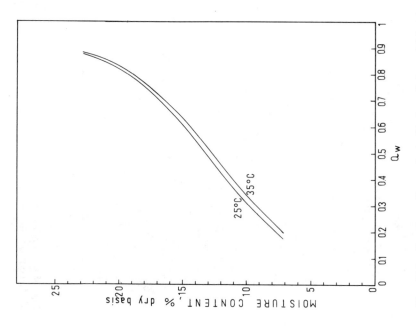

Fig. 385. Sorghum (adsorption): Variety Bukura Mahemba from Tanganyika. Method: Dew-point. (Ayerst, 1965.)

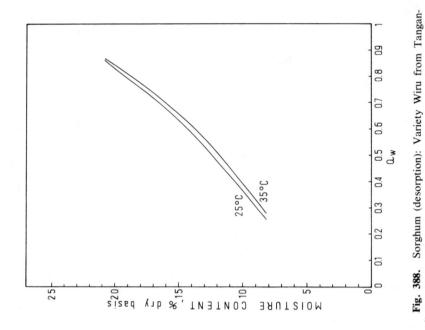

Fig. 388. Sorghum (desorption): Variety Wiru from Tanganyika. Method: Dew-point. (Ayerst, 1965.)

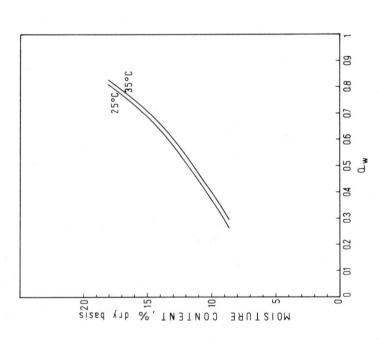

Fig. 387. Sorghum (adsorption): Variety Wiru from Tanganyika. Method: Dew-point. (Ayerst, 1965.)

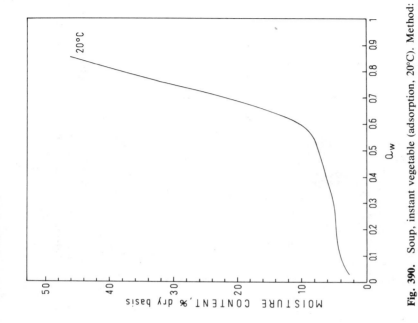

Fig. 390. Soup, instant vegetable (adsorption, 20°C). Method: Static-desiccator (saturated salt/sulfuric acid solutions). (Lewicki and Brzozowski, 1973.)

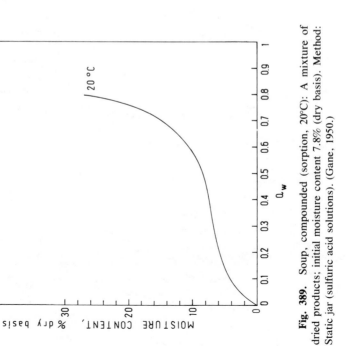

Fig. 389. Soup, compounded (sorption, 20°C): A mixture of dried products; initial moisture content 7.8% (dry basis). Method: Static jar (sulfuric acid solutions). (Gane, 1950.)

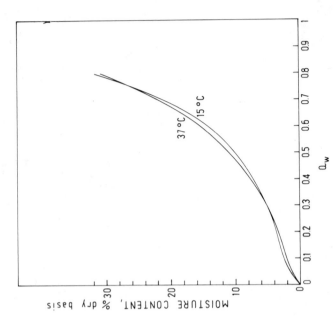

Fig. 392. Soup, vegetable, cooked (sorption): Ingredients cooked together and then roller-dried; initial moisture content 4.5% (dry basis). Method: Static-jar (sulfuric acid solutions). (Gane, 1950.)

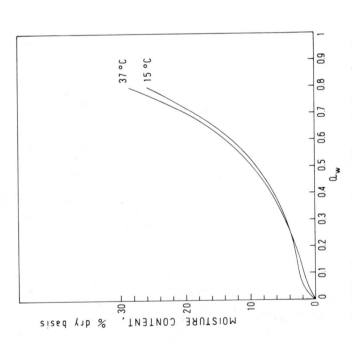

Fig. 391. Soup, meat and vegetable, cooked (sorption): Ingredients cooked together and then roller-dried; initial moisture content 3.4% (dry basis). Method: Static-jar (sulfuric acid solutions). (Gane, 1950.)

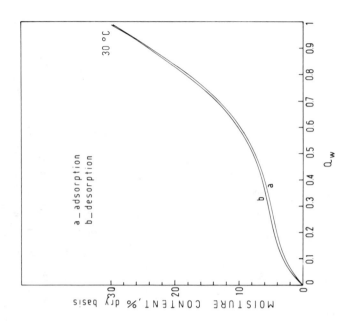

Fig. 393. Soybeans (adsorption and desorption, 30°C): Commercial samples of dry soybeans, variety Clark, with 20% oil content; desorption performed after equilibration at 75% relative humidity. Method: Vacuum sorption apparatus with quartz spring balance. (Saravacos, 1969.)

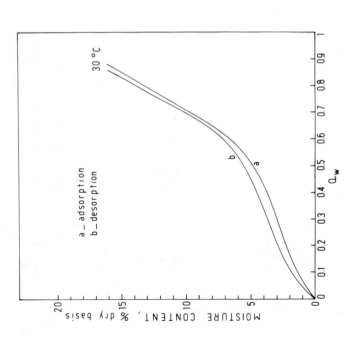

Fig. 394. Soybeans, defatted (adsorption and desorption, 30°C): Commercial samples of dry soybeans, variety Harasoy; desorption performed after equilibration at 75% relative humidity. Method: Vacuum sorption apparatus with quartz spring balance. (Saravacos, 1969.)

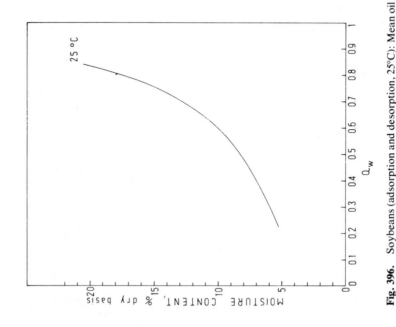

Fig. 396. Soybeans (adsorption and desorption, 25°C): Mean oil content was 20.5% (dry basis); samples for adsorption were air-dried at 30°C; for desorption, the moisture content was raised by adding water. Method: Dew-point. (Pixton and Warburton, 1971b.)

Fig. 395. Soybeans, full-fat (adsorption and desorption, 30°C): Commercial samples of dry soybeans, variety Harasoy, with 20% oil content; desorption performed after equilibration at 75% relative humidity. Method: Vacuum sorption apparatus with quartz spring balance. (Saravacos, 1969.)

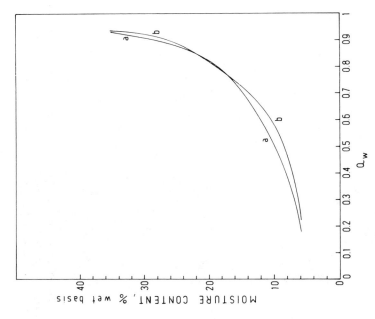

Fig. 398. Soya meal (desorption): Oil content 1.9% (dry basis); water was added to raise moisture content before desorption determinations; (a) 15, (b) 35°C. Method: Dew-point. (Pixton and Warburton, 1975a.)

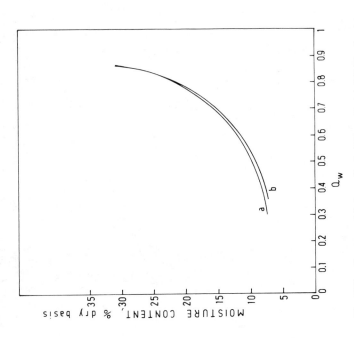

Fig. 397. Soya meal (adsorption): Oil content 1.9% (dry basis). Method: Dew-point. (Pixton and Warburton, 1975a.)

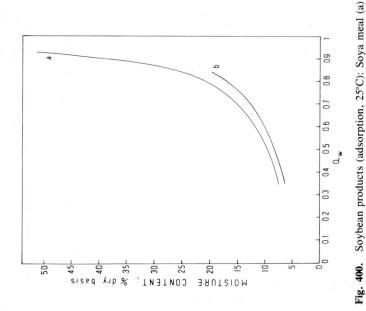

Fig. 399. Soybean products (desorption, 25°C): Soya meal (a) had 1.9% (dry basis) oil content and soybeans (b) had 20.7% (dry basis) oil content; liquid water was added to raise moisture content before desorption determinations. Method: Dew-point. (Pixton and Warburton, 1975a)

Fig. 400. Soybean products (adsorption, 25°C): Soya meal (a) had 1.9% (dry basis) oil content and soybeans (b) 20.7% (dry basis) oil content. Method: Dew-point. (Pixton and Warburton, 1975a)

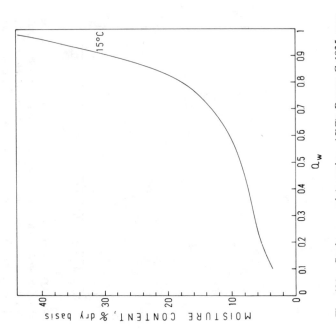

Fig. 402. Soy protein concentrate (adsorption): Air-dried at 35°C followed by vacuum-drying; material contained 69% protein, 6.1% ash, and 0.4% lipid; (a) 1, (b) 21, (c) 37°C. Method: Static-desiccator (sulfuric acid solutions). (Hansen, 1976.)

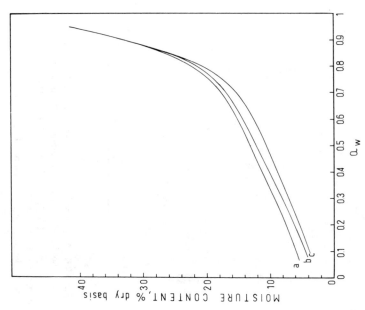

Fig. 401. Soybean seed (sorption, 15°C): Brown C 1935 crop from Dartington Hall Ltd., Totnes, Devon; initial moisture content was 9.4% (dry basis). Method: Static-jar (sulfuric acid solutions). (Gane, 1948.)

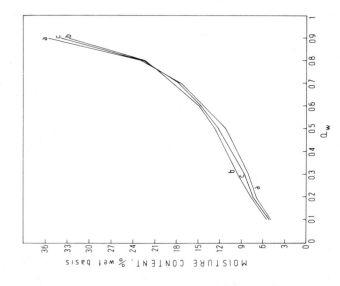

Fig. 403. Soy protein concentrate (desorption, 21°C): Air-dried at 35°C followed by vacuum-drying; material contained 69% protein, 6.1% ash, and 0.4% lipid; desorption isotherms were determined by equilibration of samples containing excess water (~ 1 g water/gram of solids). Method: Static-desiccator (sulfuric acid solutions). (Hansen, 1976.)

Fig. 404. Soy protein isolate (adsorption, 15°C): Commercially available soybean proteinate Promine-D (Central Soya): protein 86.8%, ash 4.7%; pretreatment: heating 10% dispersions for 30 min: (a): untreated, (b) pretreated at 80°C, (c) pretreated at 100°C. Method: Electrobalance assembly. (Hermansson, 1977.)

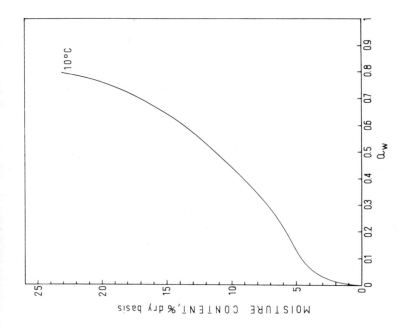

Fig. 406. Spinach (adsorption, 10°C): Scalded and freeze-dried; soluble constituents 24.3% (dry basis). Method: Static-jar (sulfuric acid solutions). (Gane, 1950.)

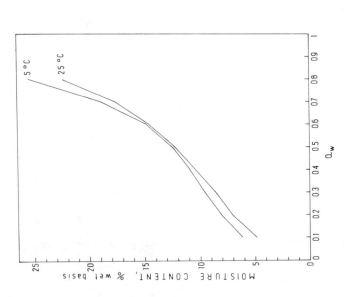

Fig. 405. Soy protein isolate (adsorption): Commercially available soybean proteinate Promine-D (Central Soya): protein 86.8%, ash 4.7%; pretreatment: heating 10% dispersions for 30 min at 80°C. Method: Electrobalance assembly. (Hermansson, 1977.)

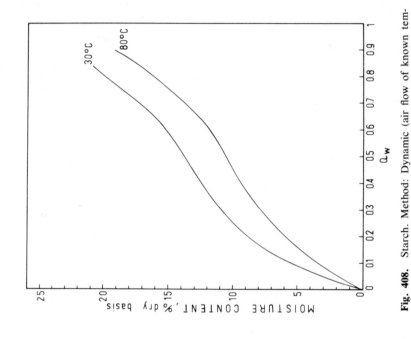

Fig. 408. Starch. Method: Dynamic (air flow of known temperature and relative humidity). (Loncin *et al.*, 1968.)

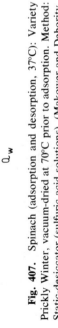

Fig. 407. Spinach (adsorption and desorption, 37°C): Variety Prickly Winter, vacuum-dried at 70°C prior to adsorption. Method: Static-desiccator (sulfuric acid solutions). (Makower and Dehority. 1943.)

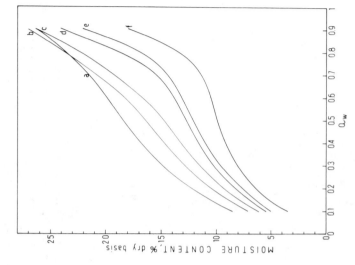

Fig. 410. Starch gel (adsorption): Freeze-dried samples at 30°C were used; before isotherm determination the samples were further vacuum-dried at 50°C; (a) − 20, − 10, 0, (b) 10, (c) 20, (d) 30, (e) 40, (f) 50°C. Method: Vacuum-sorption apparatus with quartz spring balance. (Saravacos and Stinchfield, 1965.)

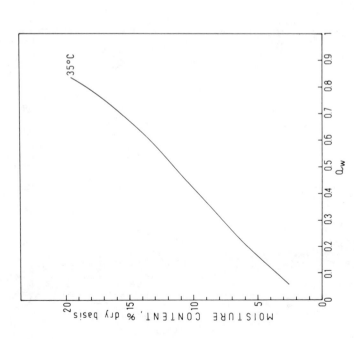

Fig. 409. Starch (adsorption, 35°C): Sample dispersed in water and freeze-dried at room temperature. Method: Static-desiccator (saturated salt/sulfuric acid solutions). (Iglesias *et al.*, 1980.)

215

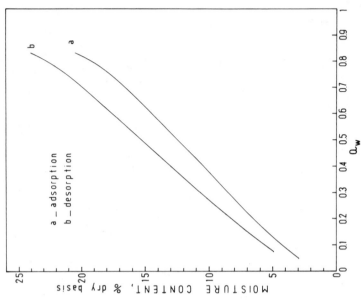

Fig. 412. Starch gel (adsorption and desorption, 35°C): A starch sample was dispersed in water with continuous heating and freeze-dried at room temperature; desorption isotherm was obtained following humidification to a_w near saturation. Method: Static-desiccator (saturated salt/sulfuric acid solutions). (Iglesias *et al.*, 1980.)

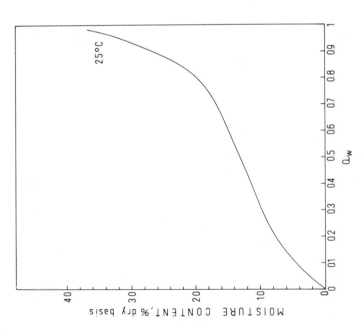

Fig. 411. Starch gel (25°C). (Fish, 1958.)

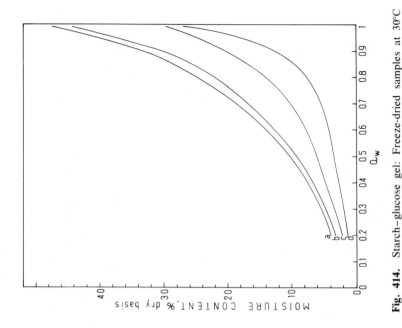

Fig. 414. Starch–glucose gel: Freeze-dried samples at 30°C were further vacuum-dried at 50°C prior to isotherm determination: (a) 30, (b) 20, (c) 40, (d) 50°C. Method: Vacuum-sorption apparatus with quartz spring balance. (Saravacos and Stinchfield, (1965.)

Fig. 413. Starch gel + Avicel (adsorption, 35°C): Starch and microcrystalline cellulose (Avicel) were wet-mixed (1:1 dry basis) with heating and freeze-dried at room temperature. Method: Static-desiccator (saturated salt/sulfuric acid solutions). (Iglesias *et al.*, 1980.)

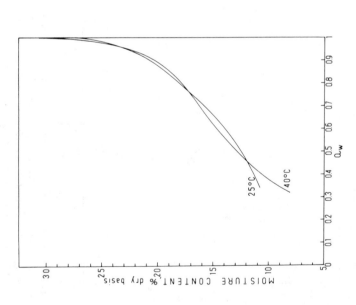

Fig. 415. Starch, maize (adsorption). Method: Dynamic (cabinet with air circulation over the samples). (Shotton and Harb, 1965.)

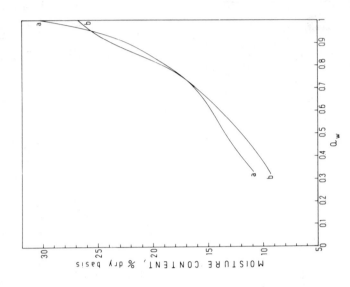

Fig. 416. Starch, maize (adsorption): (a) 30, (b) 50°C. **Method:** Dynamic (cabinet with air circulation over the samples). (Shotton and Harb, 1965.)

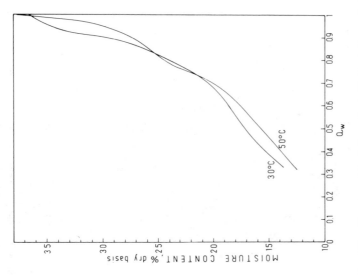

Fig. 418. Starch, potato (adsorption). **Method:** Dynamic (cabinet with air circulation over the samples). (Shotton and Harb, 1965.)

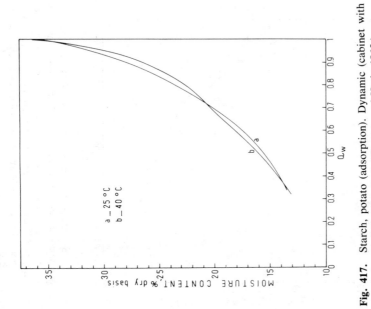

Fig. 417. Starch, potato (adsorption). Dynamic (cabinet with air circulation over the samples). (Shotton and Harb, 1965.)

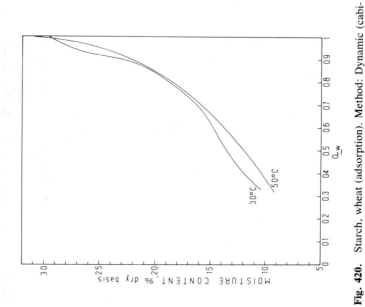

Fig. 420. Starch, wheat (adsorption). Method: Dynamic (cabinet with air circulation over the samples). (Shotton and Harb, 1965.)

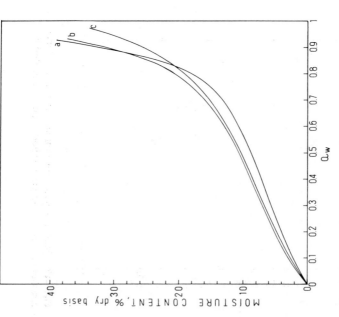

Fig. 419. Starch, potato (adsorption, 25°C): Cross-linking was carried out using phosphorus oxychloride (2 cm³/1000 g) at room temperature; pregelatinized starch was prepared on a drum drier from a slurry (40% starch): (a) pregelatinized, (b) natural, (c) cross-linked. Method: Static-jar (saturated salt solutions). (Chilton and Collison, 1974.)

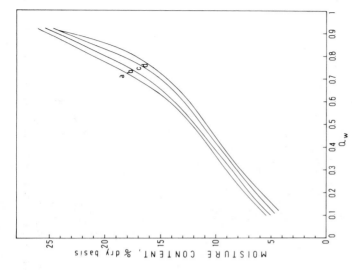

Fig. 422. Starch, wheat (adsorption): Separated from an unbleached flour, improver-free, straight-run sample commercially milled from high-grade Canadian hard red spring wheat; vacuum-dried at 65°C; (a) 20.2, (b) 30.1, (c) 40.8, (d) 50.2°C. Method: Vacuum-sorption apparatus with quartz spring balance. (Bushuk and Winkler, 1957.)

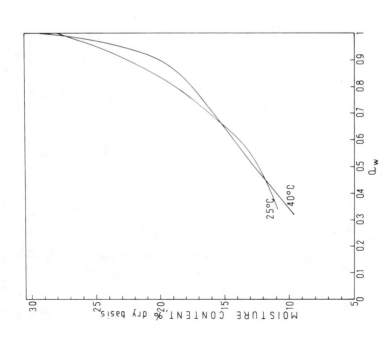

Fig. 421. Starch, wheat (adsorption). Method: Dynamic (cabinet with air circulation over the samples). (Shotton and Harb, 1965.)

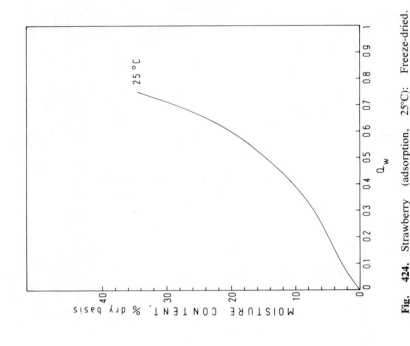

Fig. 424. Strawberry (adsorption, 25°C): Freeze-dried. Method: Static (saturated salt solutions). (Lafuente and Piñaga, 1966.)

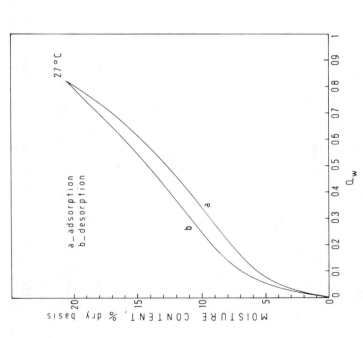

Fig. 423. Starch, wheat (adsorption and desorption, 27°C): Separated from an unbleached flour, improver-free, straight-run sample commercially milled from high-grade Canadian hard red spring wheat; vacuum-dried at 65°C. Method: Vacuum-sorption apparatus with quartz spring balance. (Bushuk and Winkler, 1957.)

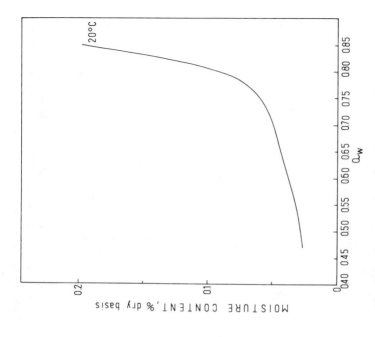

Fig. 426. Sucrose (adsorption, 20°C). Method: Static-desiccator (saturated salt/sulfuric acid solutions). (Lewicki and Brzozowski, 1973.)

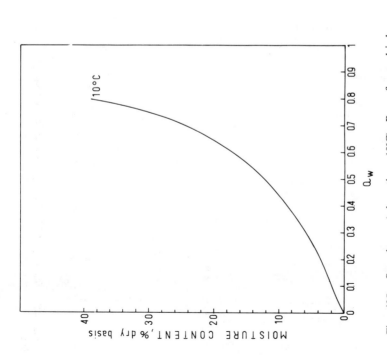

Fig. 425. Strawberry (adsorption, 10°C): Raw, freeze-dried; soluble constituents 70.4% (dry basis). Method: Static-jar (sulfuric acid solutions). (Gane, 1950.)

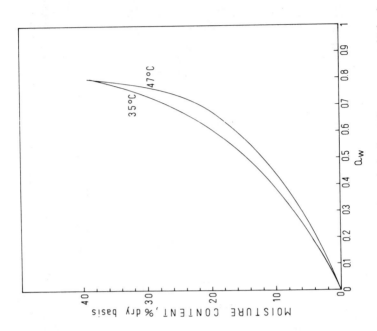

Fig. 428. Sucrose (adsorption): Amorphous sucrose obtained by freeze-drying a 20% solution. Method: Static-desiccator (saturated salt/sulfuric acid solutions). (Iglesias *et al.*, 1975b.)

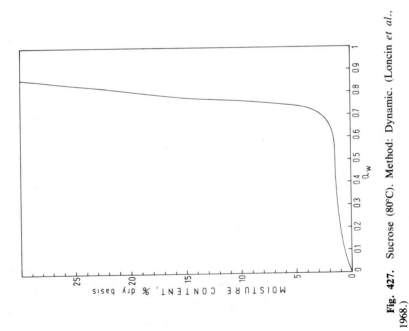

Fig. 427. Sucrose (80°C). Method: Dynamic. (Loncin *et al.*, 1968.)

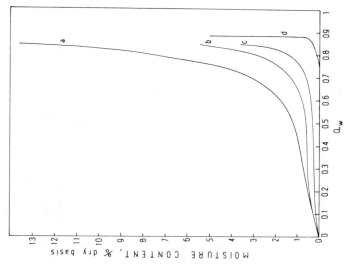

Fig. 430. Sugar, cane sugar products (sorption, 25°C): "Magma" (a) is a dark crystal mass obtained from the liquid of the first centrifugation during sugar processing; "Demerara" sugar (b) is an impurity-containing product from which refined sugar is produced; (c) unrefined, (d) amorphous sugar. Method: Static-dessiccator (saturated salt/sulfuric acid solutions). (Quast and Teixeira Neto, 1976.)

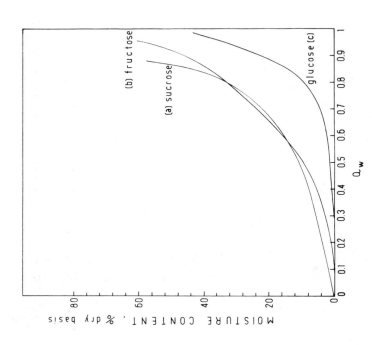

Fig. 429. Sugars (adsorption, 25°C): Amorphous sugars: (a) sucrose, (b) fructose, (c) glucose. [(a) Makower and Dye (1956); (b) Ditmar (1935); (c) Kargin (1957), as quoted by Strolle *et al.* (1970).]

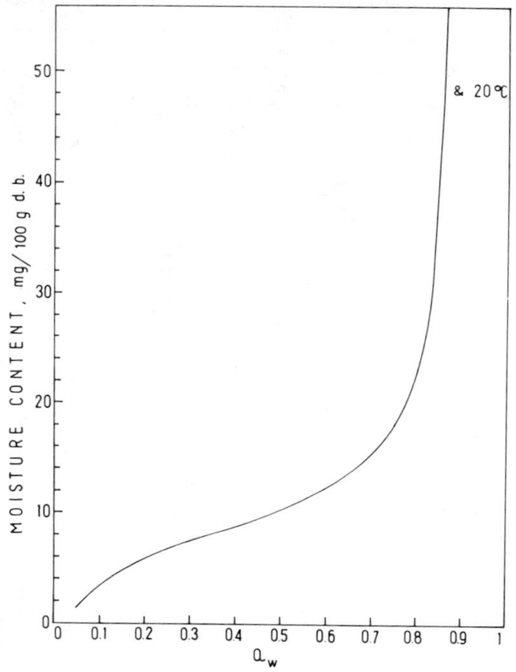

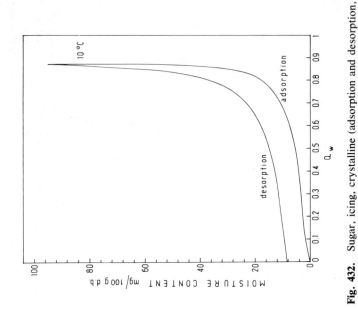

Fig. 431. Sugar, icing, crystalline (adsorption, 10, 12.5, 15, 17.5, 20°C). Method: Electrobalance assembly. (von Roth, 1977.)

Fig. 432. Sugar, icing, crystalline (adsorption and desorption, 10°C). Method: Electrobalance assembly. (von Roth, 1977.)

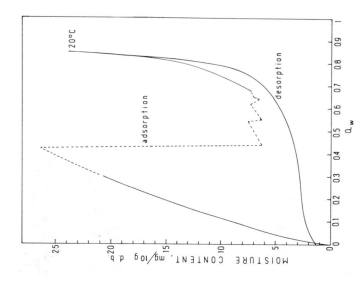

Fig. 434. Sugar, icing, freshly milled (adsorption and desorption, 20°C). First cycle. Method: Electrobalance assembly. (von Roth, 1977.)

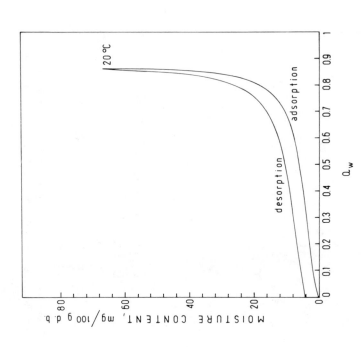

Fig. 433. Sugar, icing, crystalline (adsorption and desorption, 20°C). Method: Electrobalance assembly. (von Roth, 1977.)

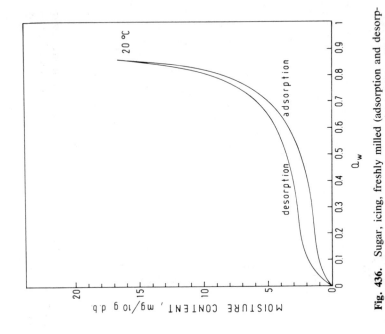

Fig. 436. Sugar, icing, freshly milled (adsorption and desorption, 20°C). Third cycle. Method: Electrobalance assembly. (von Roth, 1977.)

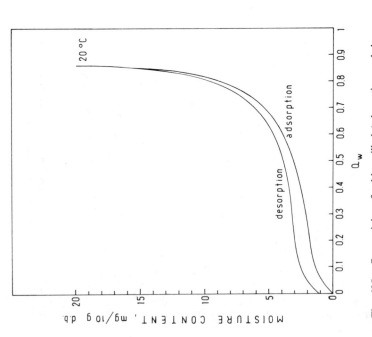

Fig. 435. Sugar, icing, freshly milled (adsorption and desorption, 20°C). Second cycle. Method: Electrobalance assembly. (von Roth, 1977.)

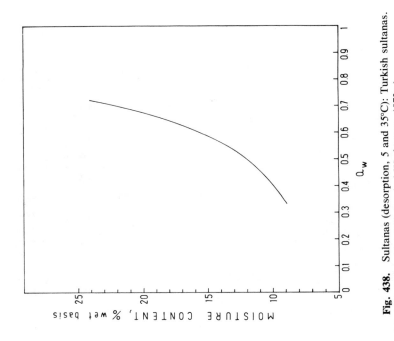

Fig. 438. Sultanas (desorption, 5 and 35°C): Turkish sultanas. Method: Dew-point. (Pixton and Warburton, 1973a.)

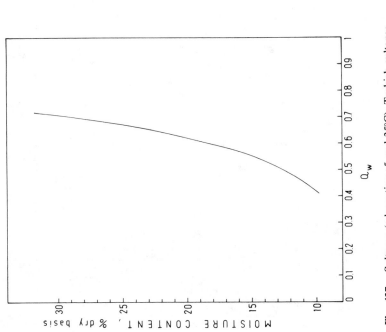

Fig. 437. Sultanas (adsorption, 5 and 35°C): Turkish sultanas. Method: Dew-point. (Pixton and Warburton, 1973a.)

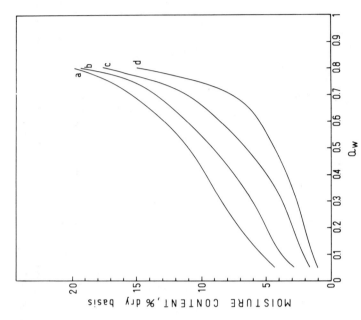

Fig. 440. Sweet marjoram (adsorption): Samples were vacuum-dried at 30°C prior to adsorption; (a) 5, (b) 25, (c) 45, (d) 60°C. Method: Jar with air agitation (sulfuric acid solutions). (Wolf *et al.*, 1973.)

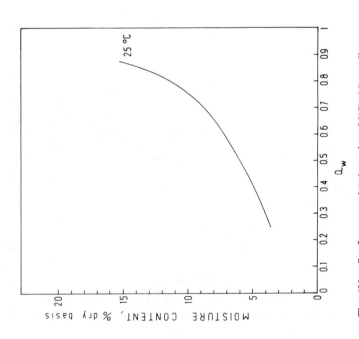

Fig. 439. Sunflower seed (adsorption, 25°C): Mean oil content 36.0% (dry basis); samples for adsorption were air-dried at 30°C. Method: Dew-point. (Pixton and Warburton, 1971b.)

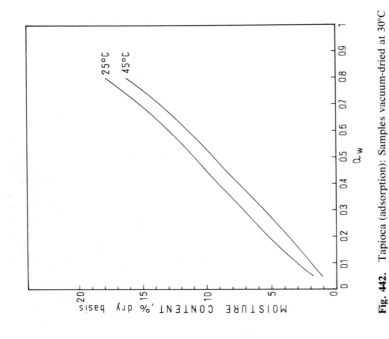

Fig. 441. Sweet marjoram (desorption): Samples for adsorption were vacuum-dried at 30°C; desorption following humidification at 90% relative humidity; (a) 5, (b) 25, (c) 45°C. Method: Jar with air agitation (sulfuric acid solutions). (Wolf *et al.*, 1973.)

Fig. 442. Tapioca (adsorption): Samples vacuum-dried at 30°C before adsorption. Method: Jar with air agitation (sulfuric acid solutions). (Wolf *et al.*, 1973.)

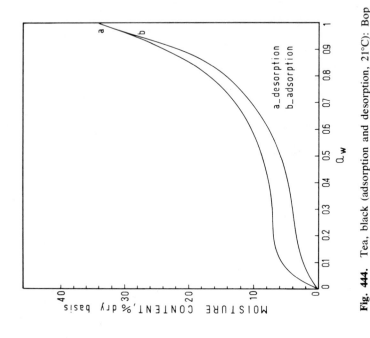

Fig. 443. Tapioca (desorption): Samples vacuum-dried at 30°C before adsorption; desorption following humidification at 90% relative humidity. Method: Jar with air agitation (sulfuric acid solutions). (Wolf *et al.*, 1973.)

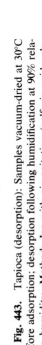

Fig. 444. Tea, black (adsorption and desorption, 21°C): Bop grade; samples dried at 70°C before adsorption determinations; for desorption samples were previously humidified to about 20%; tea size ranged between 650 and 1500 µm. Method: Static-desiccator (saturated salt solutions). (Jayaratnam and Kirtisinghe, 1974a.)

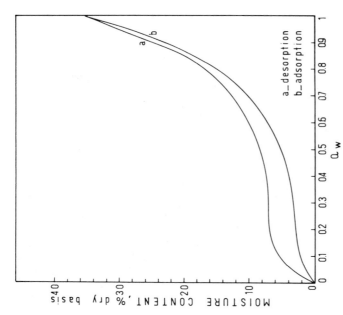

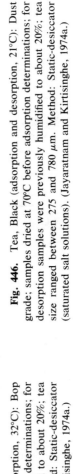

Fig. 446. Tea, Black (adsorption and desorption, 21°C): Dust grade; samples dried at 70°C before adsorption determinations; for desorption samples were previously humidified to about 20%; tea size ranged between 275 and 780 μm. Method: Static-desiccator (saturated salt solutions). (Jayaratnam and Kirtisinghe, 1974a.)

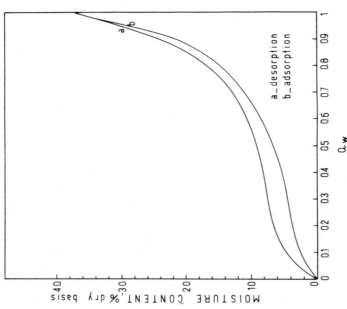

Fig. 445. Tea, black (adsorption and desorption, 32°C): Bop grade; samples dried at 70°C before adsorption determinations; for desorption samples were previously humidified to about 20%; tea size ranged between 650 and 1500 μm. Method: Static-desiccator (saturated salt solutions). (Jayaratnam and Kirtisinghe, 1974a.)

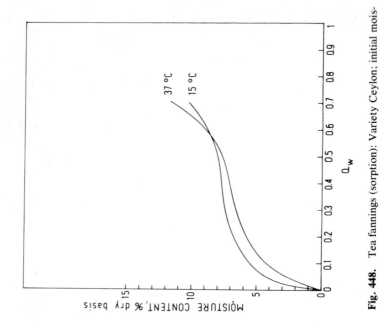

Fig. 448. Tea fannings (sorption): Variety Ceylon; initial moisture content was 8.0% (dry basis). Method: Static-jar (sulfuric acid solutions). (Gane, 1950.)

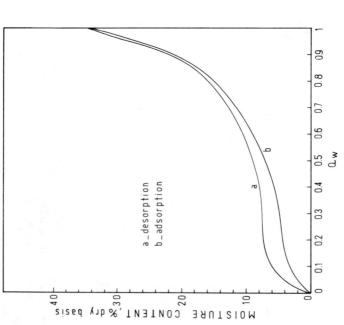

Fig. 447. Tea, black (adsorption and desorption, 32°C): Dust grade; samples dried at 70°C before adsorption determinations; for desorption samples were previously humidified to about 20%; tea size ranged between 275 and 780 μm. Method: Static-desiccator (saturated salt solutions). (Jayaratnam and Kirtisinghe, 1974a.)

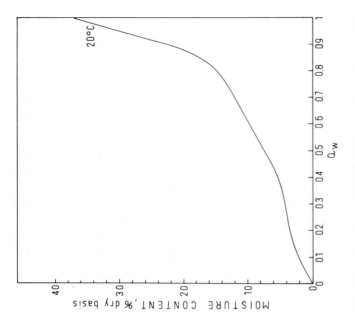

Fig. 450. Tea, leaves (20°C): High brown Broken Orange Pekoe tea. Method: Static-desiccator (saturated salt solutions). (Jayaratnam and Kirtisinghe, 1974b.)

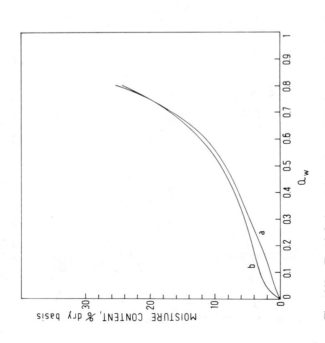

Fig. 449. Tea infusion (adsorption, 10°C): Freeze-dried; (a) Lapsang Soochong, (b) Dar Jeeling. Method: Static-jar (sulfuric acid solutions). (Gane, 1950).

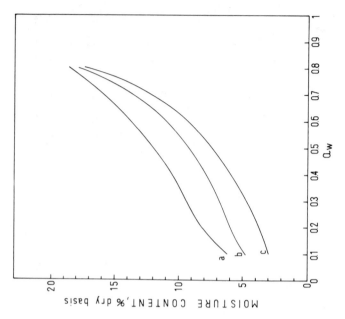

Fig. 452. Thyme (desorption): Vacuum-dried at 30°C before adsorption; desorption following humidification at 90% relative humidity; (a) 5, (b) 25, (c) 45°C. Method: Jar with air agitation (sulfuric acid solutions). (Wolf *et al.*, 1973.)

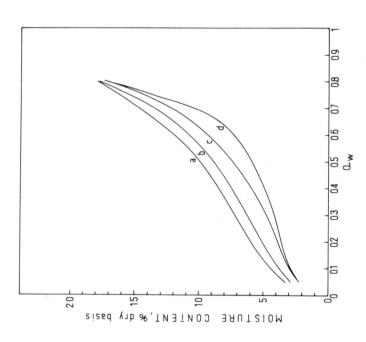

Fig. 451. Thyme (adsorption): Vacuum-dried at 30°C before adsorption; (a) 5, (b) 25, (c) 45, (d) 60°C. Method: Jar with air agitation (sulfuric acid solutions). (Wolf *et al.*, 1973.)

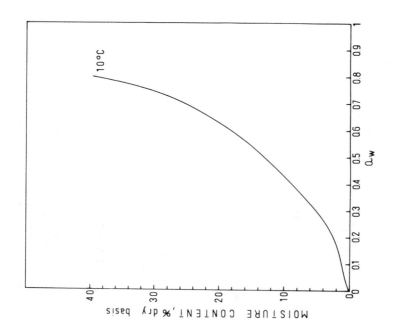

Fig. 454. Tomato (adsorption, 10°C): Raw, freeze-dried. Method: Static-jar (sulfuric acid solutions). (Gane, 1950.)

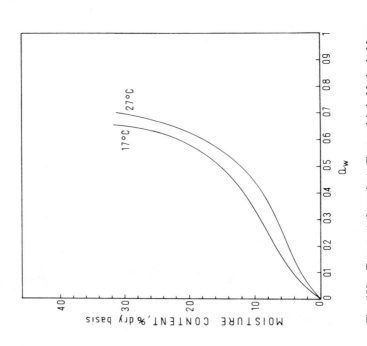

Fig. 453. Tomato (adsorption): Freeze-dried. Method: Manometric apparatus. (Alcaraz *et al.*, 1977.)

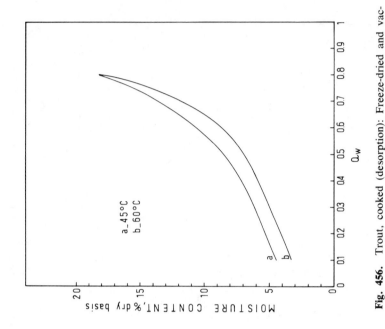

Fig. 455. Trout, cooked (adsorption): Freeze-dried and vacuum-dried at 30°C before adsorption: (a) 5, (b) 45, (c) 60°C. Method: Jar with air agitation (sulfuric acid solutions). (Wolf et al., 1973.)

Fig. 456. Trout, cooked (desorption): Freeze-dried and vacuum-dried at 30°C before adsorption; desorption following humidification at 90% relative humidity. Method: Jar with air agitation (sulfuric acid solutions). (Wolf et al., 1973.)

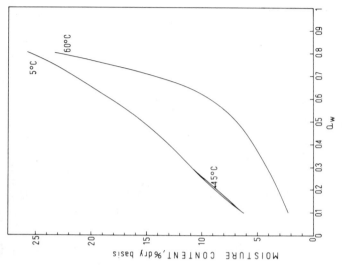

Fig. 457. Trout, raw (adsorption): Freeze-dried and vacuum-dried at 30°C before adsorption; (a) 5, (b) 45, (c) 60°C. Method: Jar with air agitation (sulfuric acid solutions). (Wolf *et al.*, 1973.)

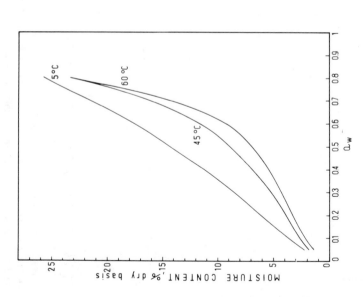

Fig. 458. Trout, raw (desorption): Freeze-dried and vacuum-dried at 30°C before adsorption; desorption following humidification at 90% relative humidity; except in the range drawn. 45°C isotherm coincides with 5°C isotherm; (a) 5, (b) 45, (c) 60°C. Method: Jar with air agitation (sulfuric acid solutions). (Wolf *et al.*, 1973.)

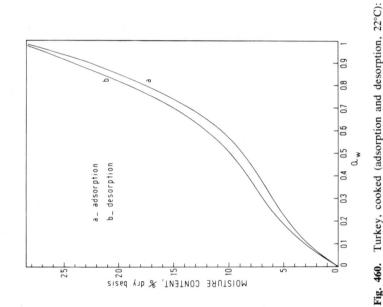

Fig. 460. Turkey, cooked (adsorption and desorption, 22°C): Breast muscle (pectoralis superficialis) of turkey precooked in boiling water and freeze-dried. Method: Vacuum-sorption apparatus with quartz spring balance. (King *et al.*, 1968.)

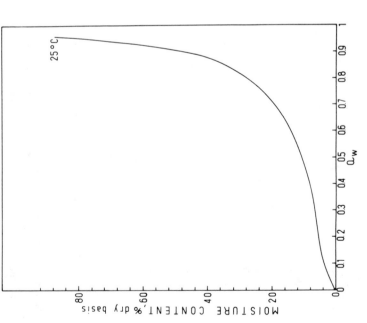

Fig. 459. Tuna, big-eye (adsorption, 25°C): The muscle of big-eye tuna (*Thunnus obesus*) was freeze-dried and ground in a mortar; total lipid content was 13.5%. Method: Static desiccator (sulfuric acid solutions). (Koizumi *et al.*, 1978.)

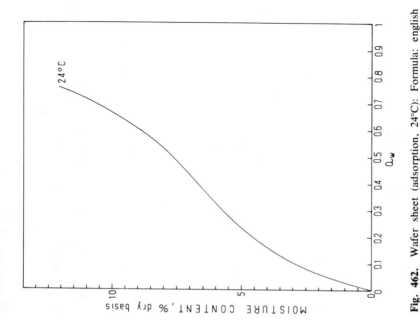

Fig. 462. Wafer sheet (adsorption, 24°C): Formula: english flour, 39.9%; water, 58.5%; powdered lecithin, 0.38%; groundnut oil, 0.95%; salt, 0.09%; NaHCO$_3$, 0.13%. Method: Dynamic (closed system with air circulation). (Barron, 1977.)

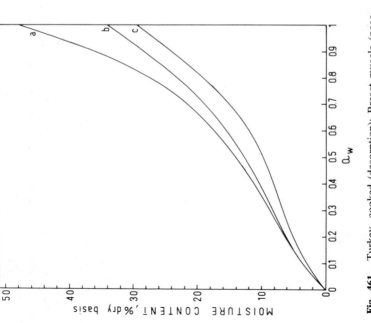

Fig. 461. Turkey, cooked (desorption): Breast muscle (pecotoralis superficialis) of turkey precooked in boiling water and freeze-dried; (a) 0, (b) 10, (c) 22°C. Method: Vacuum-sorption apparatus with quartz spring balance. (King *et al.*, 1968.)

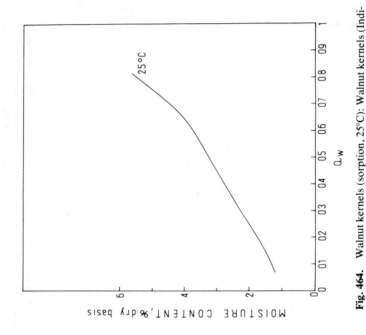

Fig. 464. Walnut kernels (sorption, 25°C): Walnut kernels (Indiana light amber halves) from Jammu, India. Method: Static (saturated salt solutions). (Vaidya *et al.*, 1977.)

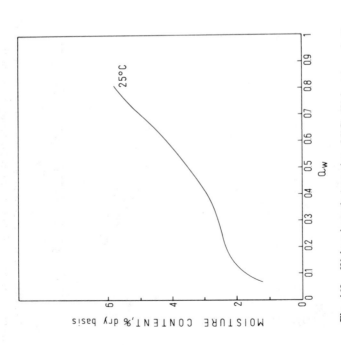

Fig. 463. Walnut kernels (sorption, 25°C): Walnut kernels (Indian brown halves) from Jammu, India. Method: Static (saturated salt solutions). (Vaidya *et al.*, 1977.)

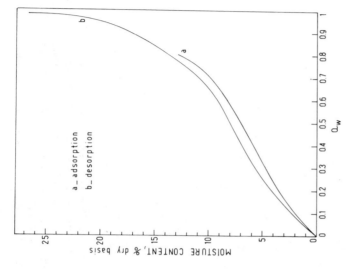

Fig. 466. Walnuts, in shell (adsorption and desorption, 22.5°C): (a) Dried bleached Placentia Perfection variety, whole in-shell nuts, 1955 crop, Ventura County, California; (b) fresh, undried, unbleached, in-shell Franquette variety, 1956 crop, Lake County, California, ground to pass $^3/_{16}$ in. orifice. Method: Dynamic. (Rockland, 1957.)

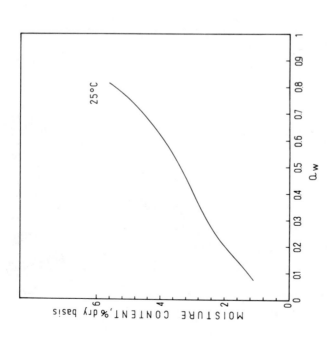

Fig. 465. Walnut kernels (sorption, 25°C): Walnut kernels (Indian light halves) from Jammu, India. Method: Static (saturated salt solutions). (Vaidya *et al.*, 1977.)

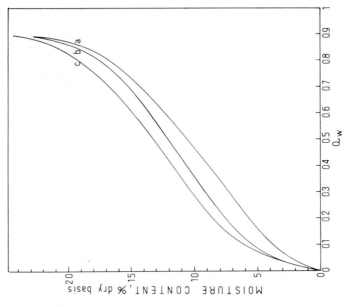

Fig. 468. Wheat (adsorption and desorption): Samples for adsorption were vacuum-dried at 72–76°C; samples for desorption were previously humidified at 97% relative humidity; values given are the average for varieties Elgin, Stewart, Pawnee, and Mida; (a) adsorption, 35°C; (b) desorption, 35°C; (c) desorption, 25°C. Method: Static-desiccator (saturated salt solutions). (Hubbard *et al.*, 1957.)

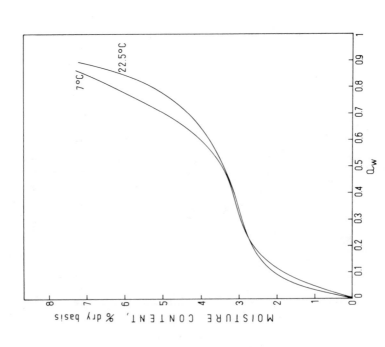

Fig. 467. Walnuts, shelled (adsorption): Placentia Perfection variety, light-colored, shelled walnut halves and pieces, 1954 crop, Ventura County, California. Method: Dynamic. (Rockland, 1957.)

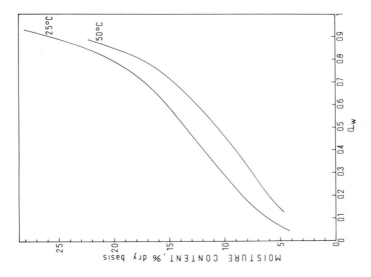

Fig. 470. Wheat (desorption): Wheat kernels tempered to an initial moisture content of 32% (dry basis). Method: Static-jar (sulfuric acid solutions). (Becker and Sallans, 1956.

Fig. 469. Wheat (adsorption and desorption, 30°C): Samples for adsorption were vacuum-dried at 72–76°C; samples for desorption were previously humidified at 97% relative humidity; values are the average for varieties Elgin, Stewart, Pawnee, and Mida. Method: Static-desiccator (saturated salt solutions). (Hubbard *et al.*, 1957.)

245

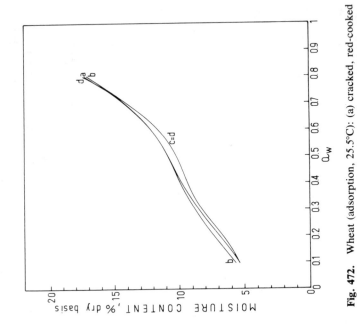

Fig. 472. Wheat (adsorption, 25.5°C): (a) cracked, red-cooked above 1 atm pressure; (b) cracked, white-cooked at 1 atm pressure; (c) cracked, red-cooked at 1 atm pressure; (d) whole kernel, red-cooked at 1 atm pressure. Method: Static-jar (saturated salt solutions). (Ferrel et al., 1966.)

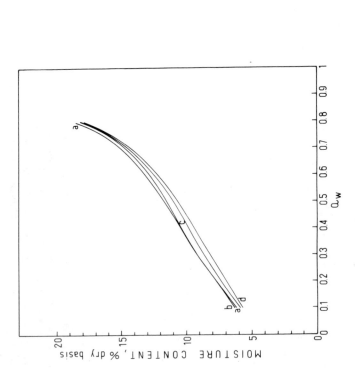

Fig. 471. Wheat (adsorption, 18.3°C): (a) cracked, red-cooked above 1 atm pressure; (b) cracked, white-cooked at 1 atm pressure; (c) cracked, red-cooked at 1 atm pressure; (d) whole kernel, red-cooked at 1 atm pressure. Method: Static-jar (saturated salt solutions). (Ferrel et al., 1966.)

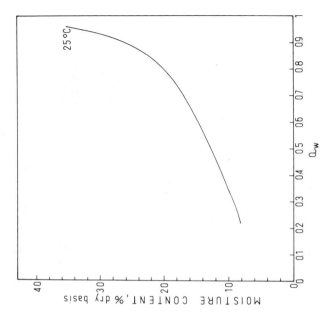

Fig. 474. Wheat, Cappelle (adsorption, 25°C). Method: Dew-point. [Pixton and Warburton (1971a) as quoted by Pixton and War-burton (1971b).]

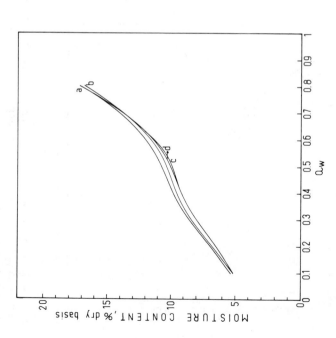

Fig. 473. Wheat (adsorption, 32.3°C): (a) cracked, red-cooked above 1 atm pressure; (b) cracked, white-cooked at 1 atm pressure; (c) cracked, red-cooked at 1 atm pressure; (d) whole kernel, red-cooked at 1 atm pressure. Method: Static-jar (saturated salt solu-tions). (Ferrel et al., 1966.)

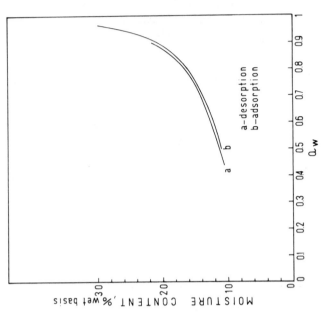

Fig. 476. Wheat, Manitoba (adsorption and desorption, 25°C). Method: Dew-point. (Averst, 1965.)

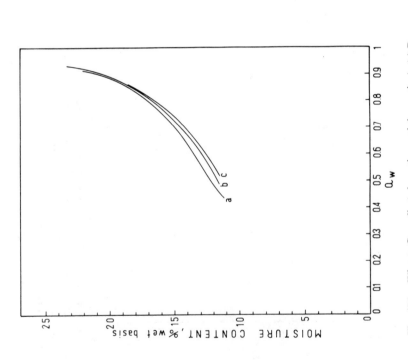

Fig. 475. Wheat, Cappelle (adsorption and desorption): (a) Desorption, 25°C; (b) adsorption, 25°C; (c) adsorption, 35°C. Method: Dew-point. (Ayerst, 1965.)

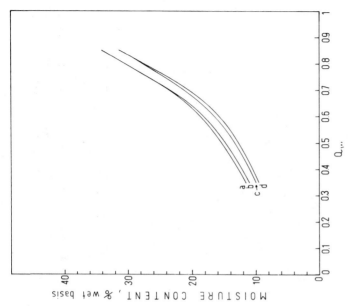

Fig. 477. Wheat, Manitoba (adsorption, 25°C). Method: Dew-point. (Ayerst, 1965.)

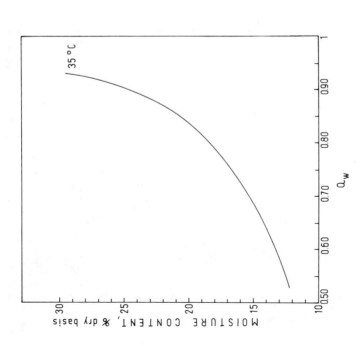

Fig. 478. Wheatfeed (adsorption and desorption): Wheatfeed was heated at 105°C for 2 hr and a 2-kg sample was mixed thoroughly with 0.4 kg glycerol; samples for adsorption were first equilibrated to a_w = 0.22; for desorption experiments the moisture content was previously raised to 26% by spraying water; adsorption: (c) 15, (d) 35°C; desorption: (a) 15, (b) 35°C. Method: Dew-point. (Warburton and Pixton, 1975.)

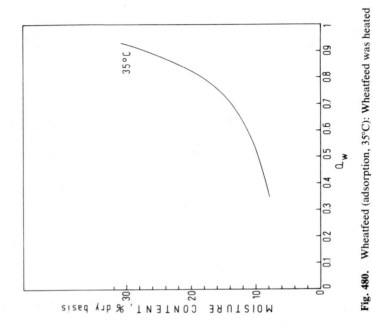

Fig. 480. Wheatfeed (adsorption, 35°C): Wheatfeed was heated at 105°C for 2 hr; samples were equilibrated to $a_w = 0.22$ prior to adsorption determinations. Method: Dew-point. (Warburton and Pixton, 1975.)

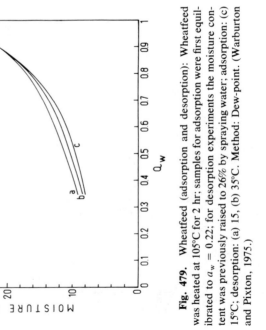

Fig. 479. Wheatfeed (adsorption and desorption): Wheatfeed was heated at 105°C for 2 hr; samples for adsorption were first equilibrated to $a_w = 0.22$; for desorption experiments the moisture content was previously raised to 26% by spraying water; adsorption: (c) 15°C; desorption: (a) 15, (b) 35°C. Method: Dew-point. (Warburton and Pixton, 1975.)

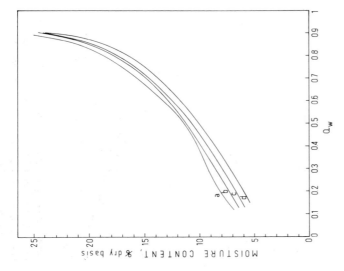
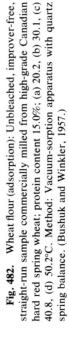

Fig. 482. Wheat flour (adsorption): Unbleached, improver-free, straight-run sample commercially milled from high-grade Canadian hard red spring wheat; protein content 15.0%; (a) 20.2, (b) 30.1, (c) 40.8, (d) 50.2°C. Method: Vacuum-sorption apparatus with quartz spring balance. (Bushuk and Winkler, 1957.)

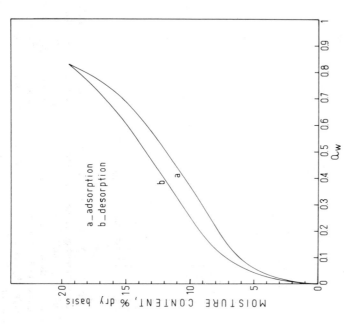

Fig. 481. Wheat flour (adsorption and desorption, 27°C): Unbleached, improver-free, straight-run sample commercially milled from high-grade Canadian hard red spring wheat; protein content 15.0%. Method: Vacuum-sorption apparatus with quartz spring balance. (Bushuk and Winkler, 1957.)

251

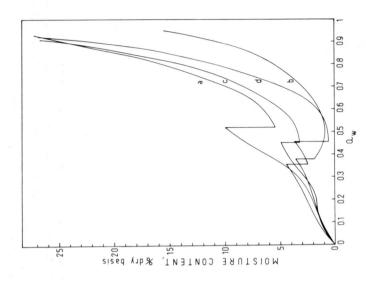

Fig. 483. Whey, Acid whey–soy beverage (sorption, 24°C): Composition of beverages (dry basis): (a) no stabilizer: whey solids, 57.0%; soy solids, 27.8%; sucrose, 13.95%; citric acid, 0.36%; aritificial sweetener, 0.82%; (b) whey solids, 56.5%; soy solids, 26.8%; sucrose, 13.15%; citric acid, 0.54%; artificial sweetener, 0.79%; carboxymethyl cellulose, 2.64%. Method: Electrobalance assembly. (Berlin *et al.*, 1973b.)

Fig. 484. Whey powders (adsorption, 24.5°C). (a) spray-dried sweet whey, (b) commercial acid whey, (c) commercial sweet whey, (d) foam-spray-dried cottage cheese whey. Method: Electrobalance assembly. (Berlin *et al.*, 1968a.)

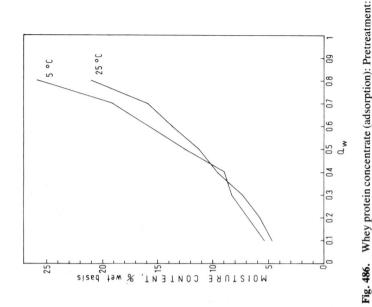

Fig. 486. Whey protein concentrate (adsorption): Pretreatment: heating a 10% dispersion for 30 min at 80°C; composition: protein, 76.8%; ash, 5.4%; fat, 5.2%; lactose, 8.3%. Method: Electrobalance assembly. (Hermansson, 1977.)

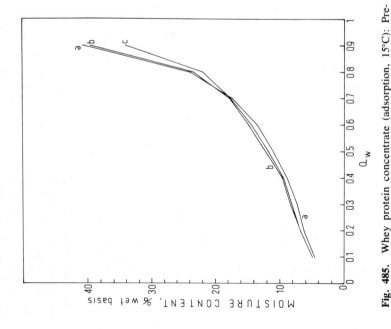

Fig. 485. Whey protein concentrate (adsorption, 15°C): Pretreatment: heating a 10% dispersion for 30 min; composition: protein, 76.8%; ash, 5.4%; fat, 5.2%; lactose, 8.3%; (a) untreated, (b) pretreated at 80°C, (c) pretreated at 100°C. Method: Electrobalance assembly. (Hermansson, 1977.)

253

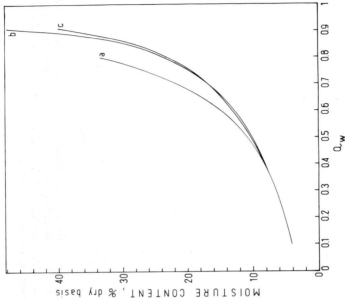

Fig. 488. Whey protein concentrate (adsorption, 24°C): Composition (dry basis): (a) (commercial product) protein, 72.8%; lactose, 7.2%; ash, 13.8%; denatured protein, 14.3% of total protein; (b) (commercial product) protein, 87.0%; lactose, 4.0%; ash, 1.6%; denatured protein, 17.6% of total protein; (c) (pilot plant product) protein, 83.0%; lactose, 5.5%; ash, 5.3%; denatured protein, 62.5% of total protein. Method: Electrobalance assembly. (Berlin et al., 1973a.)

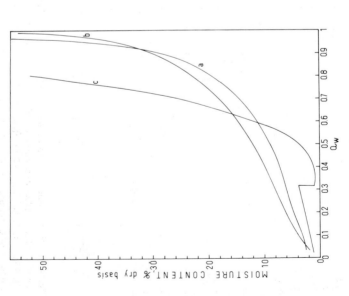

Fig. 487. Whey protein concentrate (adsorption, 24°C): Composition (dry basis): protein, 78.0%; lactose, 1.3%; ash, 12.8%; denatured protein, 29.9% of total protein: (a) whey protein concentrate, (b) nondialyzable fraction, (c) dialyzate fraction. Method: Electrobalance assembly. (Berlin et al., 1973a.)

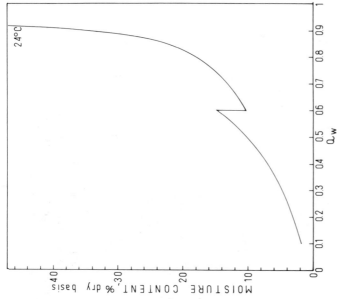

Fig. 490. Whey, sweet whey–soy beverage (sorption, 24°C): Composition of beverage: whey solids, 57.2%; soy solids, 27.3%; sucrose, 13.65%; citric acid, 1.48%; artificial sweetener, 0.63%; the same isotherm was obtained when the composition of beverage was changed: whey solids, 55.1%; soy solids, 26.8%; sucrose, 13.4%; citric acid, 1.55%; artificial sweetener, 0.62%; carboxymethyl cellulose, 1.41%; spray-dried. Method: Electrobalance assembly. (Berlin et al., 1973b.)

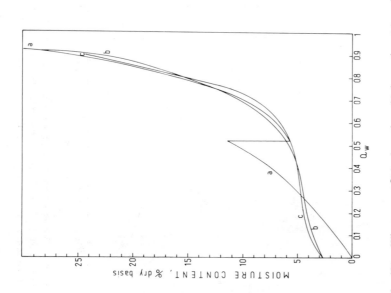

Fig. 489. Whey, sweet [(a) adsorption, (b) desorption, (c) readsorption, 24.5°C]. Foam-spray-dried. Method: Electrobalance assembly. (Berlin et al., 1968a.)

255

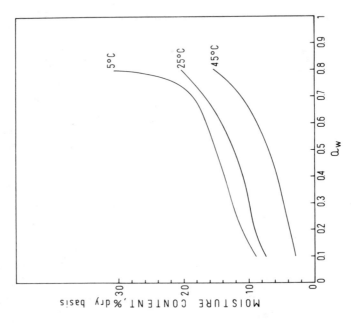

Fig. 492. Winter savory (desorption): Vacuum-dried at 30°C before adsorption; desorption following humidification at 90% relative humidity; (a) 5, (b) 25, (c) 45°C. Method: Jar with air agitation (sulfuric acid solutions). (Wolf et al., 1973.)

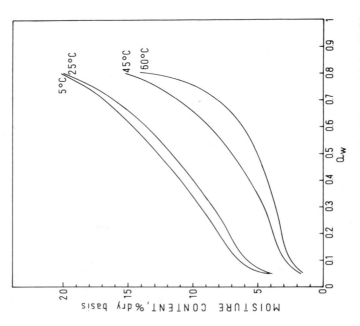

Fig. 491. Winter savory (adsorption): Vacuum-dried at 30°C before adsorption; (a) 5, (b) 25, (c) 45, (d) 60°C. Method: Jar with air agitation (sulfuric acid solutions). (Wolf et al., 1973.)

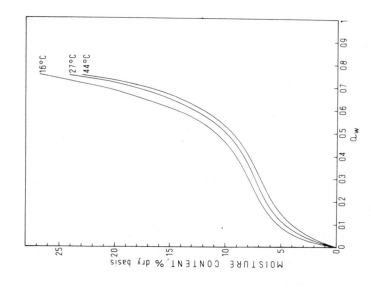

Fig. 494. Yeast, active dried (adsorption): Commercial active dried yeast; (a) 16, (b) 27, (c) 44°C. Method: Static-sealed container (saturated salt solutions). (Peri and De Cesari, 1974.)

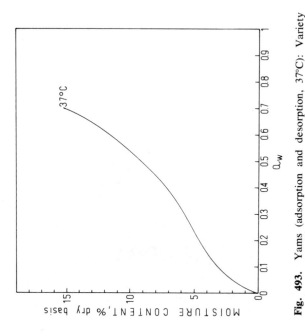

Fig. 493. Yams (adsorption and desorption, 37°C): Variety Puerto Rico. Method: Static-desiccator (sulfuric acid solutions). (Makower and Dehority, 1943.)

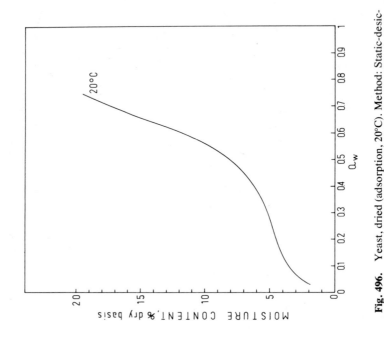

Fig. 496. Yeast, dried (adsorption, 20°C). Method: Static-desiccator (saturated salt/sulfuric acid solutions). (Lewicki and Brzozowski, 1973.)

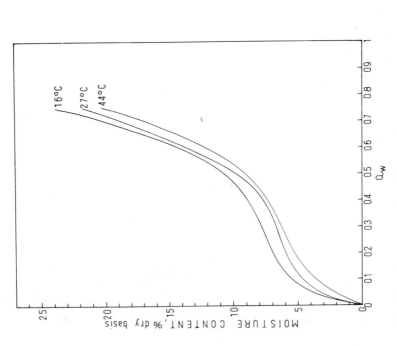

Fig. 495. Yeast, compressed (desorption): (a) 16, (b) 27, (c) 44°C. Method: Static-sealed container (saturated salt solutions). (Peri and De Cesari, 1974.

258

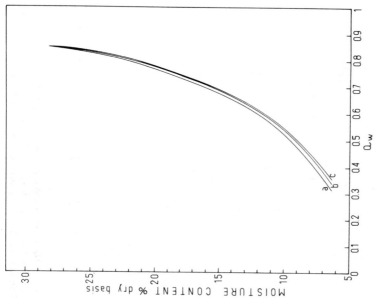

Fig. 498. Yeast, tropina (adsorption): Analysis of sample: nitrogen, 9.74%; residual hydrocarbon, 0.38%; lipid, 11.9%; ash, 7.17%; phosphorous, 1.63%; (a) 10, (b) 25, (c) 40°C. Method: Dew-point. (Pixton and Warburton, 1977b.)

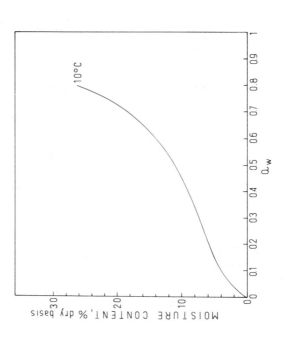

Fig. 497. Yeast (adsorption, 10°C): Freeze-dried. Method: Static-jar (sulfuric acid solutions). (Gane, 1950.)

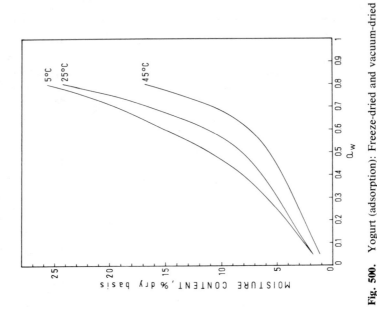

Fig. 500. Yogurt (adsorption): Freeze-dried and vacuum-dried at 30°C before adsorption; (a) 5, (b) 25, (c) 45°C. Method: Jar with air agitation (sulfuric acid solutions). (Wolf et al., 1973.)

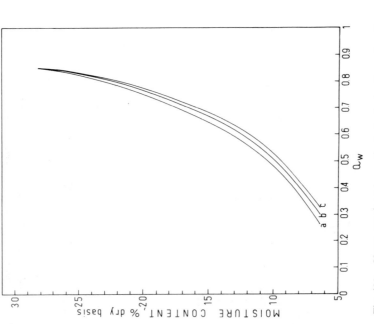

Fig. 499. Yeast, tropina (desorption): Analysis of sample: nitrogen, 9.74%; residual hydrocarbon, 0.38%; lipid, 11.9%; ash, 7.17%; phosphorus, 1.63%; (a) 10, (b) 25, (c) 40°C. Method: Dew-point. (Pixton and Warburton, 1977b.)

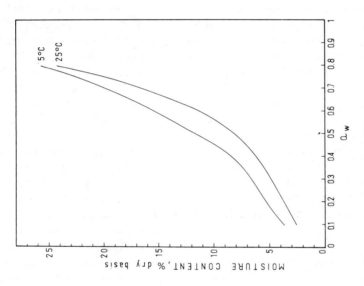

Fig. 501. Yogurt (desorption): Freeze-dried and vacuum-dried at 30°C before adsorption; desorption following humidification at 90% relative humidity. Method: Jar with air agitation (sulfuric acid solutions). (Wolf *et al.*, 1973.)

MATHEMATICAL DESCRIPTION OF
ISOTHERMS

I. INTRODUCTION

Equations for fitting water sorption isotherms in foods are of special interest in many aspects of food preservation by dehydration. Among them may be mentioned the prediction of drying times, of the shelf life of a dried product in a packaging material, or of equilibrium conditions after mixing products with various water activities. In addition to practical considerations, the isotherm equation is also needed for evaluating the thermodynamic functions of the water sorbed in foods (Iglesias *et al.*, 1976). Labuza (1968) has pointed out the need for mathematical models in order to use the isotherm with computer techniques to solve the type of problem mentioned above.

Several mathematical equations have been reported in the literature for describing water sorption isotherms of food materials. Each model— empirical, semiempirical, or theoretical—has had some success in reproducing equilibrium moisture content data. However, none of these have been able to give accurate results throughout the whole range of water activity and for different types of foods. This is mainly because moisture sorption isotherms of food products represent the integrated hygroscopic properties of numerous constituents, and the depression of water activity is due to a combination of factors, each of which may be predominant in a given range of water activity (Karel, 1973).

Chirife and Iglesias (1978) have compiled and discussed most of the isotherm equations that have been reported in the literature for fitting sorption isotherms of foods.

On the basis of the above-mentioned compilation, a statistical analysis was made on the fitting abilities of various two-parameter equations as applied to each experimental isotherm collected. This determined which equation best describes the experimental adsorption or desorption data for each situation examined in this book. The following equations were studied:

Bradley's equation (Bradley, 1936):

$$\ln \frac{1}{a_w} = B(2)B(1)^x \tag{I}$$

Halsey's equation (Halsey, 1948; Iglesias et al., 1975c):

$$a_w = \exp[-B(2)/X^{B(1)}] \tag{II}$$

Henderson's equation (Henderson, 1952):

$$1 - a_w = \exp\{-[B(2)X^{B(1)}]\} \tag{III}$$

Iglesias and Chirife's equation (Iglesias and Chirife, 1978):

$$\ln[X + (X^2 + X_{0.5})^{1/2}] = B(1)a_w + B(2) \tag{IV}$$

Iglesias and Chirife's equation (Iglesias and Chirife, 1981):

$$X = B(1)[a_w/(1 - a_w)] + B(2) \tag{V}$$

Kuhn's equation (Labuza et al., 1972):

$$X = \frac{B(1)}{\ln a_w} + B(2) \tag{VI}$$

Oswin's equation (Oswin, 1946):

$$X = B(2)[a_w/(1 - a_w)]^{B(1)} \tag{VII}$$

Smith's equation (Smith, 1947):

$$X = B(2) - B(1) \ln(1 - a_w) \tag{VIII}$$

BET equation (Brunauer et al., 1938):

$$\frac{a_w}{(1 - a_w)X} = \frac{1}{X_M C} + \frac{a_w(C - 1)}{X_M C} \tag{IX}$$

Equation (IX) was applied both to represent a particular isotherm and to calculate the parameter X_M, namely, the monolayer moisture content, since this parameter has some significance in the physicochemical stability of foods (Iglesias and Chirife, 1976c).

In Eqs. (I)–(IX), X is the moisture content expressed on a percent dry basis, with the exception of Smith's equation (VIII), which is used in either percent dry or percent wet basis. In this case, the parameters correspond with the basis (dry or wet) shown in the figure that gives the sorption isotherm.

$B(1)$ and $B(2)$ are statistical parameters to be used for the description of the isotherm, and $X_{0.5}$ [Eq. (IV)] is the moisture content at $a_w = 0.50$.

In addition, it should be noted that the above-mentioned parameters were calculated using a moisture content expressed in the same basis shown in the illustration corresponding to each isotherm fitted.

A nonlinear regression program was used for the determination of parameters $B(1)$ and $B(2)$. The details of the program are given in the Appendix. The monolayer value X_M was determined by least-squares analysis.

In order to calculate the goodness of fit of the various equations as applied to the experimental sorption data, a mean relative percentage deviation in modulus $\bar{E}(\%)$ was used:

$$\bar{E}(\%) = \frac{100}{n} \sum_{i=1}^{n} \frac{|\epsilon_i|}{X_{i(\text{observed})}},$$

where n is the number of values and $\epsilon_i = X_{(\text{predicted})} - X_{(\text{observed})}$, the absolute value. Normally, eight or more values are used to determine $\bar{E}(\%)$.

Table I shows the best equation to fit the data for each isotherm analyzed, the value of corresponding parameters, the value of $\bar{E}(\%)$, and those of the variance of regression, the latter being defined as:

$$v = \sum_{i=1}^{n} \frac{(X_{i\,\text{predicted}} - X_{i\,\text{observed}})^2}{n - 1}.$$

This statistical information is very useful when analyzing the goodness of fit of a particular equation, because in conjunction with the mean, the relative percentage deviation adequately and precisely defines the degree of success of the proposed isotherm equation. In order to standardize the presentation of the computed values of parameters $B(1)$ and $B(2)$, the number of significant figures was established in almost every case as four.

Parameters were only given for the cases where an accurate mathematical description of the isotherm was fulfilled, as can be seen by analyzing both $\bar{E}(\%)$ and v in Table I.

Table II shows the moisture content at water activity 0.50 and was included in order to completely characterize Eq. (IV).

Both tables have two ways to unequivocally locate any given isotherm: the corresponding reference as well as the figure where the isotherm is shown.

II. TABLES OF PARAMETERS FOR THE MATHEMATICAL DESCRIPTION

See Tables 1 and 2.

Table I. Mathematical Description of Food Isotherms

PRODUCT	SPECIFICATIONS	RANGE OF a_w	Eq.	B(1)	B(2)	v	E% av	X_M % d.b.	REF. FIG.
BEEF ACTOMIOSIN	21.1°C, Ads.	0.28-0.92	V	1.5779	-0.7261	0.0401	40.94	---	133 1
AGAR-AGAR	R.T.	0.10-0.80	III	1.9442	0.0018	0.1972	2.38	13.5	39 2
EGG ALBUMIN (unlyophilized)	25°C, Ads.	0.10-0.80	III	1.6670	0.0124	0.1013	2.58	6.3	24 3
	40°C, Ads.	0.10-0.80	III	1.6611	0.0138	0.0865	2.60	5.9	24 3
EGG ALBUMIN (Coagulated)	25°C, Ads.	0.10-0.80	I	0.8310	4.0081	0.0487	2.82	5.2	24 4
	40°C, Ads.	0.10-0.80	I	0.8140	4.2013	0.0081	1.40	5.0	24 4
EGG ALBUMIN (Lyophilized)	25°C, Ads.	0.10-0.80	VII	0.4222	10.0696	0.0739	3.17	5.6	24 5
	40°C, Ads.	0.10-0.80	III	1.6335	0.0165	0.0359	1.84	5.5	24 5
SERUM ALBUMIN	25°C, Ads.	0.10-0.80	VII	0.4279	11.3829	0.0297	1.36	6.5	24 6
	40°C, Ads.	0.10-0.80	III	1.7666	0.0096	0.0844	1.64	6.7	24 6

Table I (Cont.)

PRODUCT	SPECIFICATIONS	RANGE OF a_w	Eq.	B(1)	B(2)	v	E% av	X_M % d.b.	REF.	FIG.
ALGINIC ACID	25°C, Ads.	0.34-0.87	VIII	6.3201	5.8536	0.0699	1.59	---	170	7
	40°C, Ads.	0.32-0.89	VIII	6.8934	6.6155	0.0640	1.36	---	170	7
	50°C, Ads.	0.32-0.87	VIII	7.2706	5.3419	0.0512	1.24	---	170	8
CALIFORNIAN ALMONDS	15°C, Ads.	0.44-0.91	VIII	5.4411	0.4908	0.1011	3.32	---	138	9
	25°C, Ads.	0.46-0.91	VIII	5.5349	0.2918	0.0841	3.09	---	138	9
	35°C, Ads.	0.47-0.91	VIII	5.6388	0.0723	0.0787	2.88	---	138	9
	15°C, Des.	0.37-0.91	VIII	5.1008	1.2543	0.1212	3.86	---	138	10
	25°C, Des.	0.39-0.91	VIII	5.2167	0.9873	0.1329	4.18	---	138	10
	35°C, Des.	0.41-0.91	II	1.6457	9.4005	0.0030	0.32	---	138	11
MOROCCAN BITTER ALMONDS	15°C, Ads.=Des	0.40-0.93	II	1.6548	9.4275	0.0209	1.37	---	138	12
	25°C, Ads.=Des	0.42-0.93	II	1.6101	8.3748	0.0227	1.21	---	138	12
	35°C, Ads.=Des	0.45-0.93	II	1.5672	7.3376	0.0095	0.86	---	138	12
MOROCCAN SWEET ALMONDS	15°C, Ads.=Des	0.26-0.94	II	1.6586	8.2333	0.0782	2.30	---	138	13
	25°C, Ads.=Des	0.29-0.94	II	1.6180	7.2802	0.0768	2.20	---	138	13
	35°C, Ads.=Des	0.31-0.94	II	1.5807	6.5617	0.0912	2.45	---	138	13
AMYLOPECTIN	10.2°C, Ads.	0.08-0.59	III	1.7508	0.0051	0.0234	1.22	9.4	182	14
	28.2°C, Ads.	0.07-0.62	III	1.7350	0.0066	0.0193	1.16	8.4	182	14

AMYLOSE	10.2°C, Ads.	0.11-0.85	III	1.5934	0.0085	0.1239	2.00	9.3	182	15
	28.2°C, Ads.	0.07-0.61	VII	0.4946	13.8888	0.0335	1.78	7.9	182	15
ANISE	5°C, Ads.	0.10-0.80	VII	0.5155	9.0421	0.0641	3.57	4.8	190	16
	25°C, Ads.	0.10-0.80	II	1.3308	11.1189	0.0434	2.15	4.2	190	16
	45°C, Ads.	0.10-0.80	VII	0.6573	7.5054	0.0561	4.10	3.6	190	16
	5°C, Des.	0.10-0.80	II	1.7122	35.4645	0.1416	2.72	5.1	190	17
	25°C, Des.	0.10-0.80	II	1.5278	19.6380	0.0325	1.72	4.6	190	17
	45°C, Des.	0.10-0.80	VII	0.5546	8.6319	0.0767	3.38	4.5	190	17
APPLE	30°C, Ads. (a)	0.10-0.75	III	0.7537	0.1091	0.1901	7.47	---	164	20
	30°C, Ads. (b)	0.10-0.75	---	------	------	------	----	8.3	164	20
	19.5°C, Des.	0.10-0.70	II	0.7131	4.4751	0.1272	5.21	4.2	174	21
APPLE JUICE	20°C, Ads. (a)	0.10-0.60	V	10.3521	0.8906	0.0021	0.47	6.1	110	23
	20°C, Ads. (b)	0.10-0.80	VII	0.9745	10.0357	0.0630	3.27	---	110	23
APRICOT	25°C, Ads.	0.40-0.80	VI	-9.2133	0.2514	0.0317	0.76	---	122	28
ASPARAGUS	10°C, Ads.	0.10-0.80	II	1.0807	8.8163	0.1414	4.27	4.9	50	29
AVOCADO	25°C, Ads.	0.10-0.80	II	1.2531	7.0060	0.0485	4.34	3.2	190	30
	45°C, Ads.	0.10-0.80	VII	0.7386	5.2391	0.0148	3.57	2.7	190	30
	60°C, Ads.	---------	---	------	------	------	----	0.9	190	30
	25°C, Ads.	0.10-0.60	V	3.0939	0.5649	0.0040	2.93	1.6	114	31

Table I (Cont.)

PRODUCT	SPECIFICATIONS	RANGE OF a_w	Eq.	B(1)	B(2)	v	E% av	X_M % d.b.	REF.	FIG.
BANANA	25°C, Des.	0.10-0.80	II	1.4656	12.3547	0.0716	4.33	3.3	190	32
	25°C, Ads.	0.10-0.80	V	10.5514	0.1786	0.1123	6.53	4.0	190	33
	45°C, Ads.	--------	---	-------	------	------	----	3.1	190	33
	60°C, Ads.	--------	---	-------	------	------	----	2.1	190	33
	25°C, Ads.	0.10-0.80	III	0.7032	0.1268	0.0974	2.31	---	190	34
	10°C, Ads. scalded	0.10-0.80	IV	3.1200	1.5686	0.2626	2.36	5.9	50	35
BARLEY	25°C, Ads.	0.28-0.95	VII	0.3577	11.4189	0.1670	2.46	---	184	36
	25°C, Des.	0.23-0.95	VII	0.3301	12.3097	0.2954	2.76	---	184	36
	25°C, Ads.	0.10-0.80	III	2.1501	0.0031	0.1433	3.54	8.6	26	37
	25°C, Ads.	0.40-0.95	VIII	6.1556	6.6130	0.0152	0.66	---	140	38
BEANS	21°C, Ads.	0.65-0.93	III	2.0623	0.0046	0.0516	0.90	---	123	39
	25°C, Ads.	0.60-0.85	VII	0.3337	13.3695	0.0146	0.52	---	123	39
	25°C, Ads.	0.60-0.90	VIII	9.6585	3.5549	0.0119	0.36	---	123	39
	25°C, Ads. (a)	0.10-0.80	II	1.8879	77.6447	0.0695	2.10	7.0	186	40
	25°C, Ads. (b)	0.10-0.80	II	1.2952	13.4769	0.0440	2.03	4.9	107	40
RUNNER BEANS	10°C, Ads.	0.10-0.80	VII	0.7702	10.4525	0.0839	3.59	---	50	41
BEEF	20°C, Ads.	0.07-0.85	II	1.2243	7.3131	0.1371	3.74	3.3	111	42
, cooked	30°C, Ads. (a)	0.10-0.80	II	1.4400	15.9212	0.0252	1.85	4.8	72	43
, cooked	30°C, Ads. (b)	0.10-0.80	II	1.2504	8.9966	0.0396	2.83	4.2	72	43

BEEF, cooked	30°C, Ads. (c)	0.10-0.80	VII	0.6227	7.3221	0.0203	2.49	4.2	72	43
, cooked	30°C, Ads. (d)	0.10-0.80	VII	0.6520	6.7656	0.0043	1.49	4.2	72	43
, cooked	50°C, Ads.=Des	0.10-0.75	II	1.2442	9.7529	0.0685	2.68	4.6	69	44
, cooked	30°C, Ads. (a)	0.10-0.80	II	1.4767	19.8960	0.0422	2.33	5.4	69	45
, cooked	30°C, Ads. (b)	0.10-0.80	II	1.4008	15.3951	0.0243	1.79	5.1	69	45
, cooked	30°C, Ads. (c)	0.10-0.80	VII	0.5622	8.6489	0.0583	3.17	4.5	69	45
, cooked	21.1°C, Ads.	0.10-0.80	II	1.4605	21.2136	0.0761	2.62	5.9	133	46
, cooked	21.1°C, Des.	0.10-0.80	VI	-5.4402	3.8001	0.2548	3.52	6.2	133	46
, raw	21.1°C, Ads.	0.10-0.80	I	0.8911	2.9867	0.0415	2.22	---	133	47
, raw	21.1°C, Des.	--------	---	------	------	------	----	19.3	133	47
, raw	19.5°C, Des.	0.10-0.80	VII	0.5194	11.9546	0.0475	2.35	5.9	174	48
, raw	30°C, Ads.	0.10-0.80	VII	0.4292	10.8829	0.0651	2.48	6.4	166	49
, raw	40°C, Ads.	0.10-0.80	I	0.8226	4.4020	0.0746	2.62	6.2	166	49
, raw	50°C, Ads.	0.10-0.80	III	2.0365	0.0113	0.0928	4.53	5.1	166	49
BEET POWDER	25°C, Ads.	0.10-0.80	III	0.5999	0.1698	0.1143	5.59	---	95	50
BEET ROOT	5°C, Ads.	--------	---	------	------	------	----	8.7	190	52
	45°C, Ads.	--------	---	------	------	------	----	2.8	190	52
	45°C, Des.	0.10-0.80	VII	0.9018	10.3756	0.1033	6.94	---	190	51
SUGAR BEET ROOT	20°C, Des.	0.10-0.70	II	0.9645	7.5659	0.1232	2.83	5.5	65	54
	35°C, Des.	0.10-0.70	II	0.8553	5.0309	0.2730	6.65	4.9	65	54
	47°C, Des.	--------	---	------	------	------	----	5.0	65	54
	65°C, Des.	--------	---	------	------	------	----	3.9	65	54

Table I (Cont.)

PRODUCT	SPECIFICATIONS	RANGE OF a_w	Eq.	B(1)	B(2)	v	E% av	X_M % d.b.	REF.	FIG.
SUGAR BEET ROOT, (W.I.C.)	35°C, Ads.=Des	0.11-0.83	III	1.9221	0.0054	0.0510	1.83	6.9	65	55
	47°C, Ads.=Des	0.11-0.77	III	1.9093	0.0075	0.0202	1.01	6.4	65	55
BLACKCURRANT	19.5°C, Des.	0.10-0.70	II	0.9536	7.9630	0.0256	1.48	6.1	174	56
BROCCOLI	10°C, Ads.	0.10-0.80	VII	0.7423	11.0749	0.0485	3.59	5.7	50	58
CABBAGE	10°C, Ads. (raw)	0.10-0.80	VII	0.8369	12.7157	0.3087	4.27	---	50	59
	10°C, Ads. (scalded)	0.10-0.80	VII	0.7444	12.1523	0.2397	5.55	5.9	50	59
	19.5°C, Des	0.10-0.79	IV	2.8922	1.9408	0.3073	3.64	7.4	174	60
	37°C, Ads.=Des	0.10-0.70	V	8.9612	1.4053	0.0207	1.89	5.2	119	60
	37°C, Ads. (a)	0.10-0.60	V	10.2248	0.6445	0.0038	1.44	4.2	124	61
	37°C, Ads. (b)	0.10-0.60	II	0.6869	3.2302	0.0162	2.38	5.0	124	61
	0°C, Sorp.	-------	II	------	------	------	----	3.4	50	62
	10°C, Sorp.	0.10-0.80	II	1.0471	7.5091	0.0285	1.47	4.7	50	62
	25°C, Sorp.	0.10-0.80	VII	0.8352	9.5873	0.0675	4.34	9.1	50	62
	37°C, Sorp.	0.10-0.80	VII	0.8007	9.7378	0.0410	1.81	6.0	50	62
CAKE	21°C, Des.	0.22-0.92	II	1.3669	8.6989	0.0948	3.41	---	34	64

CARDAMON	5°C, Ads.	0.10-0.80	III	2.1394	0.0031	0.0707	1.93	7.3	190	65
	25°C, Ads.	0.10-0.80	II	1.8561	51.4065	0.0335	1.85	5.9	190	65
	45°C, Ads.	0.10-0.80	II	1.4406	14.9946	0.0283	2.37	4.7	190	65
	60°C, Ads.	0.10-0.80	V	3.4676	3.2649	0.1751	5.11	3.9	190	65
	5°C, Des.	0.10-0.80	III	2.4577	0.0010	0.0115	0.80	8.1	190	66
	45°C, Des.	0.10-0.80	VII	4.9129	9.1760	0.0922	3.55	5.1	190	66
β- CAROTENE - Cellulose Model	37°C, (a)	--------	---	------	------	------	----	2.9	9	67
	37°C, (b)	0.10-0.80	I	0.5216	5.0198	0.0559	8.22	2.5	9	67
CARROTS	19.5°C, Des	0.10-0.70	V	9.4131	2.3421	0.0867	3.04	4.5	174	68
	37°C, Ads. = Des. (b)	0.10-0.70	IV	3.6540	1.3393	0.0575	3.33	5.2	119	68
	70°C, Ads. = Des. (c)	0.10-0.60	IV	3.2841	1.3923	0.0432	2.52	4.1	119	68
	10°C, Ads. (a)	0.10-0.80	VII	0.8269	12.8169	0.0535	2.70	7.1	50	69
	10°C, Sorp.(b)	0.10-0.80	V	7.4925	2.2693	0.0801	3.03	4.1	50	69
	10°C, Sorp.	0.10-0.80	V	7.8444	1.9854	0.1124	3.34	4.1	50	70
	25°C, Sorp.	--------	---	------	------	------	----	4.5	50	70
	37°C, Sorp.	0.10-0.70	IV	3.6026	1.3155	0.0155	1.47	5.1	50	70
	60°C, Sorp.	0.10-0.80	IV	3.6422	1.1183	0.2215	2.29	6.2	50	70
	20°C, Ads.	0.10-0.80	V	8.5743	0.4529	0.0279	3.33	2.7	111	71

Table I (Cont.)

PRODUCT	SPECIFICATIONS	RANGE OF a_w	Eq.	B(1)	B(2)	v	E% av	X_M % d.b.	REF.	FIG.
	22°C, Ads. (a)	0.10-0.80	V	8.1036	1.0083	0.2296	3.52	3.4	112	72
	22°C, Ads. (b)	0.10-0.80	V	8.5635	1.3118	0.0974	2.68	3.1	112	72
	22°C, Ads.	0.09-0.80	VII	0.9090	10.0346	0.1122	5.63	---	112	73
	22°C, Ads.	0.09-0.80	IV	3.7766	1.1061	0.0778	4.95	5.1	112	74
	22°C, Ads. (a)	0.09-0.80	VII	0.9398	9.7108	0.0707	4.52	---	112	75
	22°C, Ads. (b)	0.10-0.80	VII	0.9079	9.4012	0.2005	3.73	---	112	75
	22°C, Ads.	--------	---	-----	------	------	----	2.4	112	76
	22°C, Ads.	0.09-0.80	VII	0.8929	10.4978	0.1594	5.34	3.9	112	77
CARROT SEEDS	10°C, Sorp.(a)	--------	---	-----	-------	------	----	6.0	49	78
	10°C, Sorp.(b)	0.10-0.70	II	2.3878	133.5877	0.0848	2.82	5.4	49	78
CASEIN	30°C,	0.10-0.80	III	2.1510	0.0044	0.1871	4.60	7.6	115	79
	80°C,	0.10-0.80	III	2.8124	0.0026	0.0593	3.54	4.9	115	79
CASHEWNUTS	25°C, Sorp.(a)	0.10-0.75	II	1.0652	2.3340	0.0175	6.66	2.2	152	83
	25°C, Sorp.(b)	0.10-0.75	III	0.8496	0.2841	0.0096	4.75	---	152	83
	25°C, Sorp.(c)	0.10-0.75	VII	0.9817	2.8657	0.0135	6.28	---	152	83
WHOLE CASHEWNUT	27°C, Ads. (a)	0.60-0.90	VIII	5.1315	3.1085	0.0089	0.61	---	131	84
	27°C, Des. (b)	0.60-0.80	VIII	4.7713	3.4302	0.0004	0.15	---	131	84
CELERY	5°C, Ads.	0.10-0.80	VII	0.6203	12.6920	0.0157	1.33	6.3	190	85
	25°C, Ads.	0.10-0.80	VII	0.6581	12.0212	0.0159	1.59	6.2	190	85

45°C, Ads.	0.10-0.80	II	0.8800	4.4222	0.1230	3.05	3.4	190	85
60°C, Ads.	----------	---	--------	-------	-------	----	3.2	190	85
45°C, Des.	0.10-0.80	II	0.8800	4.4222	0.1230	3.05	3.6	190	86

RED BLOOD CELL
CONCENTRATE

22.5°C, Ads.	----------	---	--------	-------	-------	----	8.4	35	87
22.5°C, Des.	----------	---	--------	-------	-------	----	6.5	35	87

CELLULOSE
MICROCRYSTALLINE

35°C, Ads.	0.10-0.80	VII	0.4428	5.6644	0.0821	5.34	3.1	77	88

CELLULOSE
MICROCRYSTALLINE-
OIL MODEL

37°C, Des.(a)	0.10-0.80	VII	0.4229	6.8695	0.1879	4.95	4.3	102	89
37°C, Des.(b)	0.10-0.76	III	1.6213	0.0321	0.0377	2.59	4.2	102	89
37°C, Des.(c)	0.10-0.76	VII	0.4756	5.6477	0.0140	2.79	3.7	102	89
37°C, Des.(b)	0.10-0.73	VII	0.4522	6.8538	0.1957	4.79	4.3	102	90
37°C, Des.(c)	0.10-0.73	I	0.7508	3.9645	0.0464	3.03	3.7	102	90
37°C, Ads.(a)	0.10-0.80	VII	0.3415	6.2637	0.0348	2.67	3.8	102	91
37°C, Ads.(b)	0.10-0.80	II	1.7569	17.2483	0.0521	2.75	3.2	102	91
37°C, Ads.(b)	0.10-0.80	VII	0.3955	5.8271	0.0209	2.90	3.7	102	92
37°C, Ads.(c)	0.10-0.80	VII	0.4580	5.5555	0.0462	4.30	3.5	102	92

SODIUM
CARBOXYMETHYL
CELLULOSE

24°C, Ads.	0.10-0.80	VII	0.5388	20.0598	0.1858	2.70	10.3	18	93

Table I (Cont.)

PRODUCT	SPECIFICATIONS	RANGE OF a_w	Eq.	B(1)	B(2)	v	E% av	X_M % d.b.	REF.	FIG.
CHAMOMILE TEA	5°C, Ads.	0.10-0.80	II	1.4204	21.2757	0.0498	2.21	6.2	190	94
	25°C, Ads.	0.10-0.80	II	1.4204	21.2757	0.0498	2.21	6.2	190	94
	45°C, Ads.	0.10-0.80	II	1.0367	6.1615	0.0054	1.07	4.1	190	94
	60°C, Ads.	--------	---	------	-------	------	----	2.9	190	94
	5°C, Des.	0.10-0.80	II	1.7936	71.9806	0.1832	2.98	7.2	190	95
	25°C, Des.	0.10-0.80	II	1.6541	45.2623	0.0661	1.69	6.9	190	95
	45°C, Des.	0.10-0.80	II	1.0746	6.9721	0.0114	1.63	4.0	190	95
CHEESE, (Edam)	25°C, Ads.	0.10-0.80	II	1.0668	4.9692	0.0978	3.52	3.3	190	96
	25°C, Des.	0.10-0.80	II	1.2540	8.5716	0.1507	4.31	3.5	190	97
CHEESE (Emmental)	25°C, Ads.	0.10-0.80	II	1.1889	5.9967	0.0213	2.36	3.3	190	98
	45°C, Ads.	0.10-0.80	VI	-2.5189	-0.1019	0.0508	9.06	2.2	190	98
	25°C, Des.	0.10-0.80	II	1.4435	11.9777	0.0565	3.25	3.7	190	99
CHEESE AND PARACASEIN	20°C, Des.(b)	0.10-0.80	II	2.3245	379.2907	0.2972	2.74	---	51	100
	20°C, Des.(c)	--------	---	------	--------	------	----	11.0	51	100
	20°C, Ads.(a)	0.10-0.80	II	1.8508	53.3476	0.1625	3.41	5.3	51	100
COTTAGE CHEESE WHEY	24°C, Des.(a)	--------	---	------	--------	------	----	1.9	19	101

24°C, Des.(b)	-------	---	------	------	------	----	2.8	19	101
24°C, Ads.	-------	---	------	------	------	----	5.3	19	102
CHICKEN COOKED 19.5°C, Des.	0.10-0.80	VII	0.4175	12.9971	0.0835	2.18	6.9	174	103
5°C, Ads.	0.10-0.80	III	1.5568	0.0146	0.0792	3.11	7.3	190	104
45°C, Ads.	0.10-0.80	VII	0.5386	8.5158	0.0033	0.55	4.7	190	104
60°C, Ads.	0.10-0.80	VI	-3.6178	0.9087	0.0344	3.23	3.3	190	104
5°C, Des.	0.10-0.80	III	2.2287	0.0019	0.0059	0.62	8.4	190	105
45°C, Des.	0.10-0.80	II	1.7319	32.8144	0.0173	1.11	5.0	190	105
60°C, Des.	0.10-0.80	VI	-3.4730	1.7060	0.0165	2.04	3.7	190	105
CHICKEN RAW 5°C, Ads.	-------	---	------	------	------	----	8.3	190	106
45°C, Ads.	0.10-0.80	VII	0.6503	9.6230	0.0452	3.02	5.0	190	106
60°C, Ads.	0.10-0.80	V	5.0749	2.1601	0.0665	3.79	3.7	190	106
5°C, Des.	0.10-0.80	VII	0.4223	15.5657	0.1018	1.51	8.5	190	107
45°C, Des.	0.10-0.80	II	1.2812	13.2408	0.0749	2.57	5.2	190	107
60°C, Des.	0.10-0.80	V	5.0749	2.1601	0.0665	3.79	3.8	190	107
CHIVES 25°C, Ads.	0.10-0.80	II	1.1146	11.8931	0.1365	1.99	6.1	190	108
60°C, Ads.	-------	---	------	-------	------	----	1.4	190	108
CINNAMON 5°C, Ads.	0.10-0.80	III	2.3849	0.0013	0.0855	2.13	8.0	190	109
25°C, Ads.	0.10-0.80	VII	0.3426	10.7387	0.0875	2.48	6.1	190	109
45°C, Ads.	0.10-0.80	VII	0.4337	8.1014	0.0219	2.18	4.7	190	109

Table I (Cont.)

PRODUCT	SPECIFICATIONS		RANGE OF a_w	Eq.	B(1)	B(2)	v	E% av	X_M % d.b.	REF.	FIG.
	60°C, Ads.		0.30-0.80	II	1.8015	22.9855	0.0134	1.20	---	190	109
	5°C, Des.		-------	--	-----	-----	-----	----	10.3	190	110
	25°C, Des.		0.10-0.80	III	2.2918	0.0024	0.0175	0.95	7.0	190	110
	45°C, Des.		0.10-0.80	III	1.7941	0.0122	0.0070	1.08	5.4	190	110
CITRUS JUICE	25°C, Ads.	(a)	0.11-0.75	II	1.1355	12.7186	0.1907	3.23	5.7	94	112
	25°C, Ads.	(b)	0.11-0.75	II	1.1078	10.7161	0.2004	3.21	5.4	94	112
	25°C, Ads.	(c)	0.11-0.75	II	1.0939	9.6176	0.2370	4.20	5.2	94	112
	25°C, Ads.	(d)	0.11-0.75	IV	2.8727	1.6778	0.0863	3.29	5.0	94	112
CLOVES	5°C, Ads.		0.10-0.80	VII	0.3155	8.9129	0.0582	2.31	5.0	190	113
	25°C, Ads.		0.10-0.80	II	1.7926	25.6953	0.0255	2.04	4.1	190	113
	45°C, Ads.		0.10-0.80	II	1.4351	9.9545	0.0324	2.20	3.3	190	113
	60°C, Ads.		-------	--	-----	-----	-----	----	1.6	190	113
	5°C, Des.		0.10-0.80	VII	0.2746	10.2268	0.1377	2.82	5.7	190	114
COCOA	15°C, Sorp.		-------	--	-----	-----	-----	----	3.9	50	115
	37°C, Sorp.		0.10-0.70	VI	-2.8564	2.3419	0.0339	3.19	3.6	50	115
COD	19.5°C, Des.		0.10-0.64	VII	0.4303	14.3641	0.0247	1.00	7.9	174	116
	30°C, Ads.		0.10-0.75	VII	0.5356	14.0298	0.1250	2.55	7.7	78	116
COFFEE "INKA"	20°C, Ads.		-------	--	-----	-----	-----	----	2.8	111	117
COFFEE BEANS	25°C, Ads.		0.42-0.90	VI	-2.5927	6.5013	0.1980	2.59	---	6	118

25°C, Des.	0.45-0.89	VI	-2.6207	5.6763	0.0867	1.76	---	6	118
35°C, Ads.	0.45-0.90	VI	-2.6219	6.2315	0.1755	2.46	---	6	119
35°C, Des.	0.47-0.89	VI	-2.6378	5.4453	0.1032	1.93	---	6	119
25°C, Ads.	0.55-0.85	VIII	7.8479	3.1077	0.0395	1.06	---	6	120
35°C, Ads.	0.57-0.86	VIII	8.0340	2.6911	0.0417	1.25	---	6	120
COFFEE									
15°C, Sorp.	-------	---	-----	-----	-----	----	3.1	50	121
35°C, Sorp.	0.10-0.70	V	3.5296	2.2600	0.0041	1.13	2.9	50	121
20°C, Sorp.	0.10-0.60	II	1.1738	4.4409	0.1611	7.68	2.5	56	122
30°C, Sorp.	0.10-0.60	II	0.7459	1.5689	0.0144	4.86	1.6	56	122
20°C, Sorp.	0.10-0.60	V	2.8711	3.0735	0.0746	5.14	3.2	56	123
30°C, Sorp.	0.10-0.60	V	3.2677	1.3620	0.0470	4.58	2.1	56	123
COFFEE EXTRACT									
10°C, Ads.	0.10-0.80	VI	-7.3726	-1.2415	0.0472	3.00	4.4	50	124
20°C, Ads.	0.10-0.60	VII	0.5618	10.0003	0.0623	3.49	5.0	56	125
30°C, Ads.	0.10-0.60	II	0.9143	5.0591	0.0387	2.68	4.3	56	125
20°C, Ads.	0.10-0.60	V	7.3184	2.3910	0.1151	3.71	4.5	56	126
30°C, Ads.	0.10-0.60	II	0.8036	4.0906	0.1182	6.46	3.9	56	126
20°C, Sorp.	0.10-0.60	V	7.5634	1.4461	0.1760	7.43	3.5	56	127
30°C, Sorp.	0.10-0.60	V	6.8635	1.6743	0.1877	7.36	3.5	56	127
20°C, Sorp.	0.10-0.60	V	7.8901	1.4640	0.1947	7.67	3.6	56	128
30°C, Sorp.	0.10-0.60	II	0.7887	3.4670	0.0653	6.29	3.1	56	128
COFFEE PRODUCTS 28°C, Sorp.(a)	0.10-0.77	II	0.9134	4.6227	0.0519	2.59	4.0	152	129

Table I (Cont.)

PRODUCT	SPECIFICATIONS	RANGE OF a_w	Eq.	B(1)	B(2)	v	E% av	X_M % d.b.	REF.	FIG.
	28°C, Sorp.(b)	0.10-0.80	III	0.9543	0.1256	0.0099	2.67	---	152	129
	28°C, Sorp.(c)	0.10-0.80	VII	0.6933	5.1838	0.0125	3.73	3.5	152	129
	28°C, Sorp.(d)	0.10-0.80	VII	0.7427	4.8213	0.0114	2.70	2.9	152	129
	28°C, Sorp.(e)	0.10-0.80	III	1.1644	0.1476	0.0318	3.77	2.9	152	129
COLLAGEN	25°C, Ads.	0.10-0.80	VII	0.3936	17.7244	0.0695	1.30	9.6	24	130
	40°C, Ads.	0.10-0.80	III	1.7101	0.0053	0.0792	1.80	9.5	24	130
	Ads.	---------	---	-------	------	------	-----	12.1	162	131
	Des.	---------	---	-------	------	------	-----	11.9	162	131
	Ads.	0.10-0.80	I	0.9146	5.3509	0.1393	1.89	13.8	162	132
	Des.	---------	---	-------	------	------	-----	16.9	162	132
	Ads.	---------	---	-------	------	------	-----	12.1	162	133
	Des.	---------	---	-------	------	------	-----	12.1	162	133
	Ads.	---------	---	-------	------	------	-----	16.5	162	134
	Des.	---------	---	-------	------	------	-----	19.2	162	134
SMOKED COPRA	25°C, Ads.	0.45-0.85	II	1.3003	3.5371	0.0025	0.54	----	140	136
CORIANDER	5°C, Ads.	0.10-0.80	III	1.8473	0.0115	0.0954	3.07	5.4	190	137
	25°C, Ads.	0.10-0.80	III	1.8473	0.0115	0.0954	3.07	5.4	190	137
	5°C, Des.	0.10-0.80	III	2.6170	0.0012	0.0512	2.16	7.0	190	138
	25°C, Des.	0.10-0.80	III	2.3344	0.0029	0.0440	2.13	6.1	190	138
	45°C, Des.	0.10-0.80	VII	0.4812	7.6029	0.0066	1.10	4.3	190	138

Material	Conditions	Range								
CORN	22°C, Ads.	0.10-0.80	I	0.8325	6.3766	0.0227	0.76	7.0	129	139
	50°C, Ads.	0.10-0.80	VII	0.3378	10.7490	0.0193	0.92	6.0	129	139
	30°C, Ads.	0.10-0.80	III	2.0099	0.0052	0.0106	1.14	7.0	63	140
	30°C, Des.	0.10-0.80	III	2.3104	0.0019	0.0079	0.70	7.5	63	140
	25°C, Ads. (a)	0.40-0.90	VIII	7.2829	6.2192	0.0676	1.33	---	139	141
	25°C, Ads. (b)	0.45-0.95	VIII	5.2049	6.9900	0.0094	0.59	---	139	141
CORN FLOURS	25°C, Ads. (a)	0.10-0.80	III	1.2999	0.0716	0.0366	2.43	3.5	5	142
	25°C, Ads. (b)	0.10-0.80	III	1.1809	0.1167	0.0334	4.66	2.7	5	142
	25°C, Ads. (c)	0.10-0.80	VII	0.6406	3.1498	0.0207	3.36	1.9	5	142
DEGERMED CORN FLOUR	25.5°C, Ads.	0.10-0.80	I	0.8100	12.7420	0.0163	0.93	7.9	98	143
	50°C, Ads.	0.10-0.80	I	0.7932	11.8322	0.0255	1.38	7.5	98	143
WHOLE CORN FLOUR	25.5°C. Ads.	0.10-0.80	I	0.8226	8.7432	0.0390	1.54	7.3	98	144
	50°C, Ads.	0.10-0.80	VII	0.2908	11.0956	0.0352	1.75	7.0	98	144
CORN GERM FLOUR	25.5°C, Ads.	-------	---	-----	------	------	----	6.7	98	145
	50°C, Ads.	0.10-0.80	II	2.3458	125.4631	0.0722	2.52	6.4	98	145
CORN MEAL	30°C, Ads. (a)	0.10-0.80	III	2.3530	0.0025	0.0077	0.84	6.3	179	146
	30°C, Ads. (b)	0.10-0.80	III	2.3255	0.0027	0.0083	0.86	6.2	179	146
	30°C, Des. (a)	0.10-0.80	III	2.8085	0.0007	0.0118	0.85	7.0	179	147

Table I (Cont.)

PRODUCT	SPECIFICATIONS	RANGE OF a_w	Eq.	B(1)	B(2)	v	E% av	X_M % d.b.	REF.	FIG.
	30°C, Des. (b)	0.10-0.80	III	2.7459	0.0008	0.0145	1.01	6.9	179	147
	30°C, Ads.	0.10-0.80	III	2.2810	0.0033	0.0195	1.18	6.0	179	148
	30°C, Des.	0.10-0.80	III	2.5965	0.0013	0.0415	1.59	6.7	179	148
CORN	4.5°C, Des.	0.10-0.80	I	0.8511	7.0057	0.0092	0.58	8.3	27	149
	30°C, Des.	0.10-0.80	I	0.8242	6.3686	0.0593	2.43	7.3	27	149
	50°C, Des.	0.10-0.80	I	0.8261	4.5310	0.0074	0.77	5.9	27	149
	15.5°C, Des.	0.10-0.80	I	0.8405	6.7381	0.0517	2.00	7.7	27	150
	38°C, Des.	0.10-0.80	I	0.8243	5.4559	0.0086	1.00	6.3	27	150
	60°C, Des.	0.10-0.80	I	0.7980	4.8296	0.0075	1.28	5.1	27	150
	10°C, Des.	0.14-0.80	VII	0.3244	13.4929	0.0086	0.59	7.7	52	151
	32.2°C, Des.	0.12-0.80	VII	0.3647	11.4856	0.0032	0.52	6.6	52	151
	48.9°C, Des.	0.12-0.80	VII	0.3840	10.2517	0.0242	1.53	6.0	52	151
	68.3°C, Des.	0.12-0.80	III	1.6377	0.0193	0.0359	2.11	5.1	52	151
COTTON SEED MEAL	37°C, Sorp.	0.11-0.53	VII	0.4895	8.9491	0.0142	1.95	4.8	125	152
CURD GELS	50°C, Ads. (a)	0.10-0.80	II	1.4412	15.9514	0.1735	3.71	4.3	7	153
	50°C, Ads. (b)	--------	--	-----	------	------	----	3.3	7	153
	20°C, Ads. (a)	0.16-0.83	VI	-3.9609	4.5937	0.0860	1.73	6.0	7	154
	20°C, Des. (b)	--------	--	-----	------	------	----	5.2	7	154
	20°C, Ads. (c)	0.10-0.80	V	3.6519	1.5541	0.1354	10.69	3.5	7	154
	20°C, Des. (d)	0.10-0.80	V	2.1899	1.3446	0.0366	6.39	2.9	7	154

DEXTRAN-10	23°C, Ads.	0.10-0.78	III	1.7559	0.0055	0.1288	2.03	9.4	46	155
DEXTRIN	28.2°C, Ads.	0.09-0.62	II	1.6214	32.4438	0.1009	3.97	6.2	182	156
	39.7°C, Ads.	0.07-0.33	IX	5.7300	17.4600	0.0012	0.61	5.7	182	156
	10.7°C, Ads.	0.10-0.84	II	1.7632	52.6899	0.1586	3.66	7.1	182	157
EGGS	10°C, Ads.	0.10-0.70	VII	0.4805	8.0481	0.0116	1.45	4.5	48	158
	37°C, Ads.	0.10-0.70	VII	0.5097	7.1133	0.0170	2.47	3.9	48	158
	60°C, Ads.	0.10-0.70	II	1.2875	6.4464	0.0120	2.26	3.1	48	158
	80°C, Ads.	0.10-0.70	VII	0.5893	4.6593	0.0079	1.84	2.5	48	158
DEHYDRATED EGGS	17.1°C, Ads.	0.05-0.40	I	0.7268	5.8136	0.0007	0.59	3.8	117	159
	30 C, Ads.	0.06-0.42	VII	0.4286	6.6885	0.0007	0.54	3.7	117	159
	40°C, Ads.	0.07-0.44	VII	0.4496	6.5016	0.0004	0.38	3.6	117	159
	50°C, Ads.	0.09-0.45	VII	0.4929	6.4023	0.0005	0.51	3.6	117	159
	60°C, Ads.	0.10-0.48	VII	0.5040	6.0626	0.0004	0.42	3.5	117	159
	70°C, Ads.	0.13-0.53	VII	0.5263	5.5450	0.0224	2.09	3.1	117	159
EGG PASTE	25°C, Ads.	0.10-0.80	III	1.6737	0.0177	0.0547	3.36	5.7	190	160
	45°C, Ads.	0.10-0.80	III	1.5540	0.0277	0.0205	2.36	4.9	190	160
EGG WHITE	20°C, Ads.	0.10-0.80	VII	0.4374	11.8432	0.0559	1.39	6.6	99	161
DESALTED EGG WHITE	20°C, Ads.	0.10-0.80	I	0.8339	4.3260	0.0610	2.51	5.8	99	161

Table I (Cont.)

PRODUCT	SPECIFICATIONS	RANGE OF a_w	Eq.	B(1)	B(2)	v	E% av	X_M % d.b.	REF.	FIG.
EGG WHITE AND YOLK	10°C, Ads. (a)	0.10-0.80	II	1.4833	33.4609	0.1061	2.00	7.1	48	162
EGG WHITE	10°C, Ads. (a)	0.10-0.80	II	1.4462	30.0316	0.1302	3.33	7.1	48	163
	10°C, Des. (b)	0.10-0.70	VII	0.4201	15.7572	0.0572	1.82	8.2	48	163
	10°C, Ads. (c)	0.10-0.80	VII	0.4906	12.9140	0.0240	1.01	6.8	48	163
	10°C, Des. (d)	0.10-0.70	VII	0.3715	14.7946	0.0134	0.60	8.1	48	163
WHOLE EGG	19.5°C, Des.	0.10-0.80	VII	0.4471	7.6824	0.0039	0.60	4.2	174	164
EGG YOLK	10°C, Ads. (a)	0.10-0.80	VII	0.5238	14.5079	0.0534	2.19	7.4	48	165
	10°C, Des. (b)	0.10-0.70	II	1.5809	51.0376	0.0560	1.82	8.2	48	165
	10°C, Ads. (c)	0.10-0.80	VII	0.4926	13.4118	0.0400	1.93	7.3	48	165
	10°C, Des. (d)	0.10-0.70	VII	0.3743	15.0112	0.0185	1.02	8.3	48	165
EGG PLANT	5°C, Ads.	0.10-0.80	III	1.2213	0.0197	0.1457	2.95	9.0	190	166
	25°C, Ads.	---------	---	--------	-------	-------	----	6.7	190	166
	45°C, Ads.	---------	---	--------	-------	-------	----	5.7	190	166
	60°C, Ads.	---------	---	--------	-------	-------	----	1.9	190	166
ELASTIN	25°C, Ads.	---------	---	--------	-------	-------	----	5.5	24	167
	40°C, Ads.	0.10-0.80	III	1.7118	0.0114	0.0449	2.02	5.6	24	167
FENNEL TEA	5°C, Ads.	0.10-0.80	VII	0.6610	5.6321	0.0357	3.20	2.8	190	168
	25°C, Ads.	0.10-0.80	VII	0.6610	5.6321	0.0357	3.20	2.8	190	168

45°C, Ads.	0.10-0.80	VII	0.7066	5.0395	0.0058	2.35	2.5	190	168
5°C, Des.	0.10-0.80	VII	0.4041	8.0099	0.0184	1.76	4.3	190	169
25°C, Des.	0.10-0.80	VII	0.5260	6.6910	0.0102	2.06	3.4	190	169
45°C, Des.	0.10-0.80	VII	0.7066	5.0395	0.0058	2.35	2.5	190	169
FIGS									
25°C, Ads.	0.50-0.75	VIII	19.0878	-2.5883	0.0265	0.90	---	145	170
25°C, Des.	0.40-0.75	VIII	17.0946	-6.6909	0.1171	1.98	---	145	170
ENZIMATICALLY MODIFIED MYOFIBRILLAR FISH PROTEINS									
20°C, Ads.	0.13-0.80	III	1.3239	0.0399	0.1014	4.77	4.5	97	171
20°C, Des.	0.13-0.94	III	1.6308	0.0151	0.0699	2.91	6.0	97	171
20°C, Ads.	--------	---	-------	------	-------	----	3.4	97	173
20°C, Des.	0.11-0.80	VII	0.4842	8.9026	0.0450	2.59	4.6	97	173
FISH PROTEIN CONCENTRATE									
25°C, Ads.	0.10-0.80	VII	0.3313	9.1425	0.0280	1.80	5.3	153	175
35°C, Ads.	0.10-0.80	VII	0.3389	8.5182	0.0227	1.73	5.1	153	175
42°C, Ads.	0.10-0.80	VII	0.3654	8.0828	0.0502	2.99	4.5	153	175
25°C, Ads. (a)	0.10-0.80	VII	0.3640	9.6826	0.0577	2.34	5.3	153	176
25°C, Ads. (b)	0.10-0.80	VII	0.3290	9.0895	0.0504	2.19	5.3	153	176
25°C, Ads. (c)	0.10-0.80	I	0.7687	6.6567	0.0056	0.92	4.9	153	176

Table I (Cont.)

PRODUCT	SPECIFICATIONS	RANGE OF a_w	Eq.	B(1)	B(2)	v	E% av	X_M % d.b.	REF.	FIG.
GELATIN	25°C, Ads.	0.10-0.80	VII	0.3840	15.7591	0.1116	1.92	8.5	24	179
	40°C, Ads.	0.10-0.80	I	0.8843	4.6337	0.1999	2.32	8.8	24	179
GELATIN GEL	-20°C = -10°C = = 0°C, Ads.	---------	---	------	------	-------	-----	16.3	166	180
	10°C, Ads.	---------	---	------	------	-------	-----	16.5	166	180
	20°C, Ads.	---------	---	------	------	-------	-----	13.3	166	180
	30°C, Ads.	---------	---	------	------	-------	-----	11.1	166	180
	40°C, Ads.	---------	---	------	------	-------	-----	9.0	166	180
	50°C, Ads.	---------	---	------	------	-------	-----	10.1	166	180
GELATIN	35°C, Ads.	0.20-0.80	III	1.2915	0.0349	0.1789	4.41	---	77	181
GELATIN-MICRO-CRYSTALLINE CELLULOSE	35°C, Ads.	0.20-0.83	III	1.2324	0.0594	0.0608	2.06	---	77	182
GELATIN-STARCH GEL	35°C, Ads.	0.10-0.80	III	1.3026	0.0348	0.1177	2.80	7.1	77	183
	35°C, Des.	---------	---	------	------	-------	-----	8.9	77	183
GINGER	5°C, Ads.	0.10-0.80	III	2.1334	0.0032	0.0446	1.66	7.4	190	184
	25°C, Ads.	0.10-0.80	VII	0.3506	11.4074	0.0315	1.76	7.0	190	184
	45°C, Ads.	0.10-0.80	VII	0.4628	8.3052	0.0135	0.98	4.7	190	184
	5°C, Des.	0.10-0.80	III	2.6943	0.0005	0.0394	1.26	8.2	190	185

25°C, Des.	0.10-0.80	VII	0.3123	12.0668	0.0045	0.55	6.8	190	185
45°C, Des.	0.10-0.80	VII	0.4793	8.9365	0.0515	3.05	4.7	190	185
GLUCOSE									
30°C	0.25-0.84	II	3.1122	335.1102	0.0570	2.29	---	115	186
SODIUM GLUTAMATE 20°C, Ads.									
	0.34-0.85	V	0.0727	-0.0321	0.0002	22.00	---	111	187
GLUTEN WHEAT									
20.2°C, Ads.	0.10-0.80	VII	0.3667	10.3334	0.0503	1.87	5.6	25	188
30.1°C, Ads.	0.10-0.80	VII	0.3965	9.6526	0.0822	3.28	5.6	25	188
27°C, Ads.	0.10-0.80	III	1.8073	0.0104	0.0276	2.12	6.7	25	189
27°C, Des.	0.10-0.80	III	2.3317	0.0022	0.2210	4.62	8.1	25	189
27°C, Ads.	0.10-0.80	III	1.6107	0.0231	0.0105	1.22	4.7	25	190
27°C, Des.	0.10-0.80	III	1.9568	0.0090	0.0792	3.73	6.1	25	190
GRAPEFRUIT									
5°C, Ads.	--------	---	------	------	------	----	6.5	190	191
25°C, Ads.	--------	---	------	------	------	----	6.5	190	191
45°C, Ads.	0.10-0.80	III	0.6645	0.1519	0.0436	5.28	6.5	190	191
60°C, Ads.	--------	---	------	------	------	----	1.9	190	191
5°C, Des.	--------	---	------	------	------	----	5.8	190	192
GRASS SEED									
15°C, Sorp.	0.10-0.80	II	2.4066	260.1303	0.0812	2.11	7.0	49	193
GREEN PEPPER									
22.2°C, Ads.	0.10-0.75	VII	0.7567	8.4360	0.1379	4.89	2.8	160	194
GROUNDNUTS									
25°C, Ads.	0.40-0.90	II	1.6968	10.9835	0.0087	1.05	---	140	195

Table I (Cont.)

PRODUCT	SPECIFICATIONS	RANGE OF a_w	Eq.	B(1)	B(2)	v	E% av	X_M % d.b.	REF.	FIG.
GROUNDNUT										
KERNELS	20°C, Ads.	0.37-0.89	II	1.6464	9.9880	0.0040	0.50	---	6	196
	20°C, Des.	0.34-0.92	II	1.5920	9.9093	0.0381	2.79	---	6	196
	30°C, Ads.	0.38-0.89	II	1.5996	8.9566	0.0043	0.54	---	6	197
	30°C, Des.	0.36-0.92	II	1.6117	10.2932	0.0420	2.13	---	6	197
GUAVA-TARO	22°C, Sorp. (1 : 2)	0.17-0.75	II	1.5476	23.3065	0.1032	1.86	---	130	198
	22°C, Sorp. (3 : 2)	0.17-0.75	II	1.4958	28.1557	0.0775	1.61	---	130	198
HALIBUT	25°C, Ads.	0.10-0.80	II	1.3384	19.4920	0.0777	1.52	6.3	93	199
HAZELNUT KERNELS	20°C, Ads.	0.57-0.84	VIII	3.8334	0.1305	0.0139	2.09	---	6	200
	30°C, Ads.	0.59-0.84	VIII	3.9541	-0.0656	0.0099	1.56	---	6	200
	40°C, Ads.	0.58-0.84	VIII	3.9825	-0.0712	0.0046	1.14	---	6	200
HIBISCUS TEA	25°C, Ads.	--------	----	------	------	------	----	2.2	190	201
	25°C, Des.	---------	----	------	------	------	----	3.2	190	202
HOPS	25°C, Ads.	0.08-0.74	II	1.5752	16.2079	0.0447	1.67	4.0	60	203
	25°C, Des.	0.08-0.68	II	1.3536	11.8756	0.2460	4.67	4.5	60	203
HORSE RADISH	5°C, Ads.	0.10-0.80	VII	0.4823	13.8900	0.0197	1.14	7.3	190	204

25°C, Ads.	0.10-0.80	VII	0.5039	12.9501	0.1422	3.35	6.8	190	204
45°C, Ads.	0.10-0.80	II	1.0340	6.3707	0.0633	2.67	4.5	190	204
60°C, Ads.	--------	---	--------	--------	--------	----	3.9	190	204
5°C, Des.	0.10-0.80	II	1.6670	55.3696	0.1124	2.01	6.9	190	205
45°C, Des.	0.10-0.80	II	1.0553	6.9463	0.0577	2.46	4.5	190	205
β-LACTO-GLOBULIN									
25°C, Ads.	0.10-0.80	III	1.5211	0.0166	0.1357	3.15	6.6	24	206
40°C, Ads.	0.10-0.80	III	1.5047	0.0189	0.0324	1.85	6.3	24	206
25°C, Ads.	0.10-0.80	III	1.5602	0.0167	0.0229	1.95	5.9	24	207
40°C, Ads.	0.10-0.80	III	1.4788	0.0221	0.0191	1.52	5.6	24	207
LACTOSE									
14°C, Ads.	--------	--	--------	--------	--------	----	2.1	16	209
34°C, Ads.	--------	--	--------	--------	--------	----	3.2	16	209
LAUREL									
5°C, Ads.	0.10-0.80	VII	0.3395	10.9260	0.0155	1.29	6.3	190	212
25°C, Ads.	0.10-0.80	VII	0.4905	8.4951	0.0604	2.94	4.5	190	212
45°C, Ads.	0.10-0.80	II	1.1959	6.0798	0.0059	1.54	3.1	190	212
60°C, Ads.	--------	---	--------	--------	--------	----	2.6	190	212
45°C, Des.	0.10-0.80	VII	0.5502	7.3283	0.1899	6.27	4.5	190	213
LEEK									
22°C, Ads. (a)	--------	---	--------	--------	--------	----	4.8	112	214
22°C, Ads. (b)	0.09-0.80	VII	0.7970	9.1081	0.1117	5.23	4.7	112	214
22°C, Ads. (a)	0.10-0.80	VI	-7.5189	-0.9963	0.1078	2.61	4.6	112	215

Table I (Cont.)

PRODUCT	SPECIFICATIONS	RANGE OF a_w	Eq.	B(1)	B(2)	v	E% av	X_M % d.b.	REF.	FIG.
	22°C, Ads. (b)	0.09-0.80	VII	0.7355	9.6308	0.0474	3.94	4.6	112	215
LEMON CRYSTALS	25°C, Ads.	0.10-0.60	VII	0.7386	10.7029	0.0578	3.19	5.5	96	216
LENTIL	5°C, Ads.	0.10-0.80	III	1.8757	0.0053	0.0415	1.90	7.5	190	217
	25°C, Ads.	0.10-0.80	VII	0.3962	12.2191	0.0127	0.74	6.9	190	217
	45°C, Ads.	0.10-0.80	VII	0.5246	9.2343	0.0557	2.91	5.2	190	217
	5°C, Des.	0.10-0.80	III	2.4291	0.0009	0.0285	1.07	9.3	190	218
	45°C, Des.	0.10-0.80	VII	0.5246	9.2343	0.0557	2.91	5.2	190	218
LINSEED SEED	25°C, Ads.	0.25-0.85	II	1.8218	21.1685	0.0322	2.26	---	140	219
	15°C, Sorp.	0.10-0.80	I	0.7538	5.3102	0.0987	4.35	4.0	49	220
BITTER BLUE LUPINS	15°C, Ads.	0.39-0.79	VIII	7.5289	4.1890	0.0070	0.62	---	138	221
	25°C, Ads.	0.42-0.80	VIII	7.6146	3.7974	0.0031	0.41	---	138	221
	35°C, Ads.	0.44-0.80	VIII	7.7531	3.4283	0.0065	0.59	---	138	221
	15°C, Des.	0.31-0.88	II	1.8685	64.0621	0.1540	1.53	---	138	222
	25°C, Des.	0.35-0.88	II	1.8169	51.9598	0.1359	1.48	---	138	222
	35°C, Des.	0.38-0.89	II	1.7688	42.9004	0.0932	1.12	---	138	222
MACARONI	15°C, Ads.=Des	0.20-0.80	III	2.1264	0.0036	0.0058	0.59	---	142	223
	25°C, Ads.=Des	0.25-0.80	III	2.0243	0.0052	0.0063	0.62	---	142	223
	35°C, Ads.=Des	0.30-0.80	VIII	5.5278	5.4145	0.0544	2.09	---	142	224

25°C, Ads. (a)	0.60-0.80	VIII	4.4697	7.4201	0.0004	0.13	---	142	225
25°C, Ads. (b)	0.65-0.80	VIII	3.0832	8.3287	0.0013	0.18	---	142	225
MALTOSE 23°C, Ads.	0.10-0.76	IV	2.9364	1.7159	0.1526	2.60	5.8	46	226
MARROW 10°C, Ads.	0.10-0.80	VII	0.7212	12.3433	0.0359	2.13	7.4	50	228
DRIED MILK BABY FOOD 25°C, Des.	0.20-0.75	V	2.0524	3.3974	0.0324	2.73	---	180	229
25°C, Des.	0.20-0.74	II	1.8793	20.8827	0.0545	2.80	---	180	230
MILK 24.5°C, Ads. (a)	--------	---	------	-------	-------	---	3.3	14	231
24.5°C, Ads. (b)	---------	---	------	-------	-------	---	3.8	14	231
24.5°C, Ads. (c)	---------	---	------	-------	-------	---	4.1	14	231
NON-FAT DRY MILK 30°C, Ads.	---------	---	------	-------	-------	---	6.7	57	232
37.8°C, Ads.	0.10-0.80	VII	0.4571	8.5704	0.1211	5.14	6.2	57	232
1.7°C, Des.	---------	---	------	-------	-------	---	7.7	57	233
15.5°C, Des.	---------	---	------	-------	-------	---	7.0	57	233
30°C, Des.	0.10-0.80	II	1.9684	72.3080	0.1419	3.83	6.5	57	233

Table I (Cont.)

PRODUCT	SPECIFICATIONS	RANGE OF a_w	Eq.	B(1)	B(2)	v	E% av	X_M % d.b.	REF.	FIG.
	37.8°C, Des.	0.10-0.80	II	1.9927	67.8072	0.1476	3.83	6.1	57	233
	30°C, Ads.	-------	---	----	----	----	----	7.2	57	234
	37.8°C, Ads.	---------	---	-------	------	----	----	6.2	57	234
	1.7°C, Des.	---------	---	-------	------	----	----	8.6	57	235
	15.5°C, Des.	---------	---	-------	------	----	----	7.6	57	235
	30°C, Des.	---------	---	-------	------	----	----	6.9	57	235
	37.8°C, Des.	---------	---	-------	------	----	----	6.2	57	235
	15.5°C, Ads.	---------	---	-------	------	----	----	7.8	57	236
	30°C, Ads.	---------	---	-------	------	----	----	6.4	57	236
	37.8°C, Ads.	0.10-0.80	VII	0.4392	8.2078	0.0892	3.84	5.2	57	236
	1.7°C, Des.	---------	---	-------	------	----	----	8.9	57	237
	15.5°C, Des.	---------	---	-------	------	----	----	7.7	57	237
	30°C, Des.	---------	---	-------	------	----	----	6.9	57	237
	37.8°C, Des.	---------	---	-------	------	----	----	6.0	57	237
MILK POWDER COMPONENTS	24.5°C, Ads. (a)	0.10-0.80	III	1.9485	0.0056	0.0498	1.24	7.0	15	243
	24.5°C, Ads. (b)	---------	---	-------	------	----	----	7.6	15	243
SKIMMILK POWDER	20°C, Ads.	0.10-0.80	VI	-3.0113	1.3983	0.0191	2.28	2.8	111	244
	14°C, Ads.	---------	---	-------	------	----	----	6.1	16	245

24°C, Ads.	---	---	---	---	---	---	4.9	16	245
34°C, Ads.	---	---	---	---	---	---	5.5	16	245
34°C, Ads. (a)	---	---	---	---	---	---	5.1	16	246
34°C, Des. (b)	0.10 0.80	II	2.0544	54.3870	0.0218	1.57	4.7	16	246
34°C, Ads. (c)	0.10-0.80	II	1.7764	23.8439	0.0190	1.90	4.0	16	246
14°C, Ads. (a)	---	---	---	---	---	---	4.1	16	247
14°C, Des. (b)	0.10-0.80	II	2.6527	290.2579	0.0145	0.92	---	16	247
14°C, Ads. (c)	0.10-0.80	II	2.3658	124.8257	0.0068	0.72	5.0	16	247
24.5°C, Ads. (a)	0.08-0.79	VII	0.4053	9.9337	0.0407	2.09	5.6	15	248
24.5°C, Ads. (b)	---	---	---	---	---	---	5.5	15	248
WHOLE MILK									
24°C, Ads.	---	---	---	---	---	---	3.1	16	249
34°C, Ads.	---	---	---	---	---	---	3.5	16	249
24.5°C, Ads. (a)	---	---	---	---	---	---	3.3	14	250
24.5°C. Des. (b)	0.10-0.80	VI	-1.4503	3.1356	0.0288	2.96	3.1	14	250
24.5°C, Ads. (c)	0.10-0.80	II	2.1884	37.9004	0.0149	1.62	3.5	14	250
MUSCLE									
7°C, Ads.	0.15-0.89	VII	0.4803	13.2452	0.2319	2.22	8.3	183	251

Table I (Cont.)

PRODUCT	SPECIFICATIONS	RANGE OF a_w	Eq.	B(1)	B(2)	v	E% av	X_M % d.b.	REF.	FIG.
MUSCLE FIBERS	Ads. (a)	0.10-0.85	VII	0.7267	8.3130	0.0364	2.16	4.9	116	252
	Ads. (b)	0.10-0.85	II	1.1948	9.4235	0.1496	5.14	4.6	116	252
	Ads. (c)	0.10-0.85	II	1.3205	14.1236	0.0973	3.06	4.9	116	252
MUSHROOMS	25°C, Ads.	--------	--	------	-------	------	----	6.4	107	253
	20°C, Ads.	0.10-0.80	V	8.3477	2.2506	0.2013	5.12	4.7	64	253
	20°C, Ads.	0.07-0.75	II	1.1639	7.5335	0.1069	4.48	3.4	111	254
	5°C, Ads.	--------	--	------	-------	------	----	4.1	190	255
	25°C, Ads.	--------	--	------	-------	------	----	4.1	190	255
	45°C, Ads.	0.10-0.80	VII	0.8702	8.0644	0.0136	2.60	4.3	190	255
	60°C, Ads.	--------	--	------	-------	------	----	2.9	190	255
	5°C, Des.	0.10-0.80	VII	0.6239	11.7399	0.0682	3.08	5.6	190	256
	25°C, Des.	0.10-0.80	VII	0.6928	10.7062	0.0790	3.04	5.4	190	256
	25°C, Ads.	--------	--	------	-------	------	----	4.9	190	257
	25°C, Des.	0.10-0.80	II	1.1606	11.5342	0.2700	3.55	5.0	190	257
	10°C, Ads.	--------	--	------	-------	------	----	6.9	50	258
MUSHROOM LIQUID EXTRACT	25°C, Ads.	--------	--	------	-------	------	----	11.9	11	259
MYOSIN A	7°C, Ads.	--------	--	------	-------	------	----	8.8	183	260
	7°C, Des.	0.12-0.85	II	1.5138	43.0877	0.0769	2.11	8.8	183	260
MYOSIN B	7°C, Ads.	0.15-0.78	VII	0.4210	12.3632	0.2009	2.75	6.8	183	261

NOODLES	7°C, Des.	0.12-0.79	VII	0.3721	15.6188	0.0979	2.03	8.5	183	261
	7°C, Ads. (b)	--------	---	------	-------	-----	----	14.5	183	262
	20°C, Ads.	0.07-0.85	VII	0.3430	9.9875	0.0305	1.78	5.5	111	263
JAPANESE NOODLES	20°C, Ads.	0.10-0.80	III	2.2850	0.0027	0.1343	3.85	7.6	169	264
	20°C, Des.	0.10-0.80	III	2.6498	0.0009	0.0655	2.28	7.6	169	264
	20°C, Ads.	0.10-0.80	III	2.2102	0.0033	0.0575	2.53	7.1	169	265
	20°C, Des.	0.10-0.80	III	2.5708	0.0011	0.0382	1.60	7.6	169	265
	20°C, Ads.	0.10-0.80	III	2.1770	0.0036	0.0609	2.45	7.3	169	266
	20°C, Des.	0.10-0.80	III	2.4710	0.0014	0.0348	1.63	7.6	169	266
	20°C, Ads.	0.10-0.80	III	2.0106	0.0052	0.1000	2.25	7.0	169	267
	20°C, Des.	0.10-0.80	III	2.2959	0.0022	0.1529	2.58	7.6	169	267
	20°C, Ads.	0.10-0.80	III	1.9789	0.0055	0.1347	2.17	7.0	169	268
	20°C, Ads.	0.10-0.80	VII	0.3356	12.0074	0.0712	2.33	7.3	169	268
	20°C, Ads.	0.10-0.80	VII	0.3792	11.1561	0.0803	2.63	6.8	169	269
	20°C, Des.	0.10-0.80	VII	0.3434	12.0674	0.0960	2.48	7.2	169	269
NUTMEG	5°C, Ads.	0.10-0.80	III	2.5585	0.0023	0.0176	1.32	5.4	190	270
	25°C, Ads.	0.10-0.80	VII	0.3473	7.8596	0.0029	0.67	4.5	190	270
	45°C, Ads.	0.10-0.80	VII	0.4841	6.3978	0.0040	1.25	3.7	190	270
	60°C, Ads.	0.10-0.80	II	1.3113	5.3696	0.1093	5.18	2.7	190	270
	5°C, Des.	0.10-0.80	III	3.1630	0.0004	0.0169	1.18	6.5	190	271

Table I (Cont.)

PRODUCT	SPECIFICATIONS	RANGE OF a_w	Eq.	B(1)	B(2)	v	E% av	X_M % d.b.	REF.	FIG.
	25°C, Des.	0.10-0.80	VII	0.3182	8.2460	0.0062	0.70	4.7	190	271
	45°C, Des.	0.10-0.80	VII	0.4276	6.8533	0.0127	1.50	3.9	190	271
PEANUT OIL	30°C	0.21-0.76	VI	-0.0359	0.0468	0.0000	2.58	---	115	272
	80°C	0.21-0.85	VI	-0.0416	0.0774	0.0001	5.56	---	115	272
OLEIC ACID	30°C	0.20-0.89	I	0.0203	2.7871	0.0000	1.24	---	115	273
	80°C	0.20-0.80	I	0.0699	2.9265	0.0005	1.97	---	115	273
ONION	10°C, Ads.	----------	---	------	------	------	----	6.5	121	274
	30°C, Ads.	----------	---	------	------	------	----	4.5	121	274
	45°C, Ads.	----------	---	------	------	------	----	3.4	121	274
	10°C, Des.	----------	---	------	------	------	----	7.2	121	275
	30°C, Des.	----------	---	------	------	------	----	7.6	121	275
	17°C, Ads.	0.10-0.70	VII	0.7070	16.1993	0.0254	1.37	9.4	3	277
	27°C, Ads.	0.10-0.70	VII	0.7923	13.5043	0.0159	1.48	9.5	3	277
ONION SEED	10°C, Sorp.	0.10-0.80	VII	0.2744	11.4643	0.0842	1.48	6.4	49	279
ORANGE CRYSTALS	25°C, Ads. (a)	0.06-0.75	III	0.7871	0.0876	0.1769	9.97	1.1	85	280
	37°C, Ads. (b)	0.07-0.71	VI	-11.7633	-3.2407	0.1194	6.29	---	85	280
	25°C, Ads. (c)	----------	---	------	------	------	----	1.1	85	280
ORGEAT	25°C, Ads.	0.10-0.80	VII	0.4785	7.9278	0.0855	2.45	4.5	136	282

	Conditions	Range	Type							Page
OVALBUMIN	21°C, Ads.	0.16-0.80	II	1.6214	31.0665	0.1191	2.31	---	54	283
PAPAYO-TARO	22°C, Sorp. (1 : 2)	0.17-0.75	II	1.6401	34.9202	0.1879	2.85	---	130	284
	22°C, Sorp. (3 : 2)	0.17-0.75	II	1.4797	35.3323	0.1446	1.71	---	130	284
PARANUT	5°C, Ads.	0.10-0.80	VII	0.3745	4.0105	0.0096	2.26	2.2	190	285
	25°C, Ads.	0.10-0.80	II	1.6822	5.1712	0.0045	1.65	1.8	190	285
	60°C, Ads.	0.10-0.80	V	1.3425	0.5983	0.0164	5.82	1.1	190	285
	5°C, Des.	0.10-0.80	VII	0.2950	4.4279	0.0036	1.07	2.5	190	286
	60°C, Des.	0.10-0.80	V	1.2464	1.0094	0.0106	3.75	1.2	190	286
DRIED PARSLEY	20°C, Ads.	--------	---	------	------	------	----	3.5	111	287
PARSNIP SEED	10°C, Sorp.	0.10-0.80	VII	0.2733	9.3314	0.0375	1.96	5.0	49	288
PEA FLOUR	20°C, Ads.	0.07-0.85	II	1.9305	49.6628	0.1218	4.44	5.5	111	289
PEACH	23.9°C, Ads. 20°C = 30°C	0.10-0.80	III	0.5777	0.2131	0.0083	5.30	---	167	290
	Ads.	0.10-0.80	III	1.0096	0.0471	0.1170	3.59	8.7	166	291
	40°C, Ads.	0.10-0.80	III	1.1909	0.0440	0.0060	0.92	6.9	166	291
	50°C, Ads.	0.10-0.80	III	1.3371	0.0477	0.0228	3.67	5.5	166	291
PEANUTS	25°C, Des. (a)	0.11-0.86	II	1.9237	31.9494	0.0614	3.10	4.3	89	292

Table I (Cont.)

PRODUCT	SPECIFICATIONS	RANGE OF a_w	Eq.	B(1)	B(2)	v	E% av	X_M % d.b.	REF.	FIG.
	25°C, Des. (b)	0.11-0.86	II	2.0636	28.7049	0.0483	2.58	3.5	89	292
	25°C, Des. (c)	0.11-0.86	III	2.0259	0.0046	0.0837	1.83	6.9	89	292
PEANUT KERNEL	10°C, Des.	0.24-0.87	II	2.6132	91.9583	0.0130	0.95	---	12	293
	21.1°C, Des.	0.23-0.87	II	2.1314	29.5992	0.0161	1.26	---	12	293
	32.2°C, Des.	0.22-0.76	VII	0.3438	5.5980	0.0095	1.39	---	12	293
	15°C, Ads.	0.10-0.80	II	1.2540	4.3079	0.0267	5.13	2.8	191	294
PEANUT SHELL	21.1°C, Des.	0.23-0.75	III	1.7936	0.0074	0.0893	2.22	---	12	295
	32.2°C, Des.	0.22-0.86	III	1.7131	0.0107	0.0617	1.23	---	12	295
PEANUT WHOLE POD	10°C, Des.	0.24-0.87	II	2.0461	47.2213	0.0565	2.50	---	12	296
	21.1°C, Des.	0.23-0.87	VII	0.3656	7.4725	0.0332	1.23	---	12	296
	32.2°C, Des.	0.22-0.76	II	1.7839	20.3363	0.0148	1.35	---	12	296
PEAR	25°C, Ads.	0.10-0.80	IV	3.5844	1.6324	0.0717	2.10	9.7	190	297
	25°C, Des.	0.10-0.80	IV	3.5844	1.6324	0.0717	2.10	9.7	190	297
	25°C, Ads.	0.10-0.80	III	0.7654	0.0882	0.1123	3.31	16.8	190	298
	25°C, Des.	0.20-0.80	IV	3.5613	1.6337	0.0107	0.44	11.0	190	298
PEAS	10°C, Ads.	0.10-0.80	II	1.0963	8.0019	0.1019	4.25	5.0	50	299
	19.5°C, Des.	0.10-0.80	VI	-4.8054	3.1117	0.1250	1.67	5.0	174	300
	25°C, Ads.	0.10-0.80	II	1.2113	10.5023	0.0618	2.89	4.4	107	300

DRIED GREEN PEAS 15°C, Ads.	0.23-0.91	VIII	7.2775	6.0261	0.0459	0.89	---	138	301
25°C, Ads.	0.25-0.91	VIII	7.2502	5.7178	0.0453	1.04	---	138	301
35°C, Ads.	0.28-0.92	VIII	7.2251	5.4163	0.0329	0.94	---	138	301
15°C, Des.	0.37-0.91	VIII	6.9995	6.8754	0.0435	1.03	---	138	302
25°C, Des.	0.40-0.91	VIII	7.0132	6.3784	0.0381	0.85	---	138	302
35°C, Des.	0.44-0.92	VIII	7.0522	5.9235	0.0356	0.90	---	138	302
PEAS SEEDS 10°C, Sorp.(a)	0.10-0.80	VII	0.2993	13.5040	0.1381	2.25	8.2	49	303
10°C, Sorp.(b)	0.10-0.80	II	2.2901	227.0831	0.0864	2.15	7.3	49	303
PEKANUT 5°C, Ads.	0.10-0.80	VII	0.4457	3.5025	0.0053	1.58	1.9	190	304
25°C, Ads.	0.10-0.80	VII	0.4457	3.5025	0.0053	1.58	1.9	190	304
45°C, Ads.	0.10-0.80	VII	0.5689	2.7634	0.0091	2.44	1.6	190	304
60°C, Ads.	0.10-0.80	II	0.8620	0.9791	0.0312	12.42	0.8	190	304
5°C, Des.	0.10-0.80	II	2.0370	10.5285	0.0265	3.57	2.0	190	305
45°C, Des.	0.10-0.80	II	1.7871	5.9190	0.0193	2.28	1.7	190	305
60°C, Des.	0.10-0.80	V	1.2365	0.5471	0.0260	4.88	0.9	190	305
PEPPERMINT TEA 5°C, Ads.	0.10-0.80	VII	0.4127	11.5537	0.1002	2.34	6.8	190	306
25°C, Ads.	0.10-0.80	VII	0.4127	11.5537	0.1002	2.34	6.8	190	306
45°C, Ads.	0.10-0.80	II	1.3011	10.8261	0.0393	3.04	4.7	190	306
60°C, Ads.	---------	---	------	------	------	---	3.2	190	306
5°C, Des.	0.10-0.80	VII	0.2913	13.8223	0.0150	0.69	7.9	190	307
25°C, Des.	0.10-0.80	II	2.1598	159.0035	0.0288	1.43	7.1	190	307

Table I (Cont.)

PRODUCT	SPECIFICATIONS	RANGE OF a_w	Eq.	B(1)	B(2)	v	E% av	X_M % d.b.	REF.	FIG.
	45°C, Des.	0.10-0.80	II	1.3837	13.6307	0.0676	2.83	4.5	190	307
PINEAPPLE	5°C, Ads.	0.10-0.80	III	0.8096	0.0821	0.1107	4.27	19.5	190	309
	25°C, Ads.	0.10-0.80	III	0.8096	0.0821	0.1107	4.27	19.5	190	309
	45°C, Ads.	0.10-0.80	III	0.7314	0.1123	0.0854	5.09	10.8	190	309
	60°C, Ads.	0.10-0.80	IV	3.9218	1.1142	0.1552	2.77	7.3	190	309
PISTACHO NUT KERNEL	20°C, Des.	0.54-0.89	VII	0.5246	9.7059	0.0856	1.39	---	36	311
PISTACHO NUT SHELL	20°C, Ads.	0.41-0.95	II	2.0388	36.8805	0.0598	1.66	---	36	312
PLUM JUICE	20°C, Ads.	0.10-0.80	III	0.6579	0.1488	0.1938	7.61	---	110	315
PORK	4.4°C, Ads.	0.10-0.80	III	1.4024	0.0523	0.0212	3.16	4.5	189	316
	4.4°C, Des.	0.19-0.83	II	2.0618	38.9852	0.0481	2.45	4.2	189	316
	19.5°C, Des.	0.10-0.78	VII	0.4455	13.5089	0.0612	2.10	7.0	174	317
POTATO	19.5°C, Des.	0.10-0.80	VII	0.3740	13.6643	0.1404	2.45	7.6	174	318
	25°C, Ads.=Des	0.10-0.70	VII	0.5975	9.8173	0.0523	3.23	5.2	64	318
	37°C, Ads.=Des	0.10-0.70	VII	0.4430	10.9452	0.0210	1.33	5.9	119	318
	30°C, Ads. (a)	0.10-0.77	VII	0.6053	7.3095	0.0320	3.43	4.7	164	319
	30°C, Ads. (b)	0.10-0.77	III	1.3444	0.0355	0.0231	1.53	5.7	164	319

30°C, Ads. (c)	0.10-0.76	III	1.5009	0.0165	0.0158	1.12	7.3	164	319
37°C, Sorp.	-------	---	------	------	------	----	5.1	40	320
20°C, Ads.	0.20-0.80	III	1.8290	0.0080	0.0271	1.40	---	166	321
30°C, Ads.	0.20-0.80	VII	0.4080	9.6726	0.0345	1.66	---	166	321
40°C, Ads.	0.20-0.80	III	1.8745	0.0132	0.0923	2.68	---	166	321
50°C, Ads.	0.20-0.80	I	0.7240	6.6095	0.0923	4.19	---	166	321
10°C, Sorp.	0.10-0.80	II	1.5068	27.4606	0.2271	2.59	6.3	50	322
37°C, Sorp.	0.10-0.80	II	1.6622	33.2730	0.1063	3.33	5.9	50	322
60°C, Sorp.	-------	---	------	------	------	----	5.4	50	322
80°C, Sorp.	0.10-0.80	II	1.5638	17.1200	0.0710	3.86	4.7	50	322
10°C, Ads. raw	0.10-0.80	VII	0.4989	11.7740	0.0087	0.98	6.5	50	323
10°C, Ads. mashed	0.10-0.80	III	1.7477	0.0076	0.0779	2.25	7.6	50	323
15°C, Sorp.	0.10-0.80	II	1.5158	22.4896	0.1371	3.33	4.9	50	324
15°C, Sorp.	0.10-0.70	II	1.8259	58.3943	0.0407	1.99	5.9	50	325
37°C, Sorp.	0.10-0.70	II	1.7666	41.6712	0.0618	2.51	5.1	50	325
28°C, Sorp.	0.10-0.70	V	4.6844	5.7366	0.0547	2.09	5.6	50	326
50°C, Sorp.	0.10-0.70	VI	-4.7444	3.7985	0.0354	1.20	5.4	50	326
28°C, Sorp.	0.10-0.70	VII	0.3076	9.4895	0.0472	2.06	5.5	50	327
50°C, Sorp.	0.10-0.70	VII	0.3788	9.1876	0.0283	1.86	5.2	50	327
70°C, Sorp.	0.10-0.70	I	0.8051	4.3115	0.0118	1.58	5.4	50	327
28°C, Sorp.	0.10-0.70	II	2.0135	74.0611	0.0332	1.88	5.8	50	328

Table I (Cont.)

PRODUCT	SPECIFICATIONS	RANGE OF a_w	Eq.	B(1)	B(2)	v	E% av	X_M % d.b.	REF.	FIG.
	50°C, Sorp.	0.10-0.70	II	1.6608	30.4792	0.0307	2.00	5.5	50	328
	70°C, Sorp.	0.10-0.70	VII	0.4704	8.8881	0.0129	1.81	5.1	50	328
POTATO FLAKES	20°C, Ads.	0.07-0.85	II	1.7490	37.6598	0.0652	2.25	5.5	111	329
	25°C, Ads.	0.12-0.76	II	1.4714	18.1936	0.1570	4.20	4.8	173	330
PRUNES	23.9°C, Ads.	0.10-0.80	III	0.8661	0.0701	0.1814	4.11	12.6	167	331
	10°C, Sorp."C"	-------	---	------	------	------	----	2.9	50	332
	10°C, Sorp."D"	-------	---	------	------	------	----	3.3	50	332
α and β-PSEUDO-GLOBULIN	25°C, Ads.	0.10-0.80	III	1.7200	0.0085	0.1109	2.54	7.2	24	333
	40°C, Ads.	0.10-0.80	VII	0.4154	11.7798	0.0336	1.81	7.0	24	333
γ-PSEUDO GLOBULIN	25°C, Ads.	0.10-0.80	III	1.7103	0.0089	0.1146	2.76	7.3	24	334
	40°C, Ads.	0.10-0.80	III	1.7178	0.0094	0.0172	1.05	7.0	24	334
RADISH	5°C, Ads.	---------	---	------	------	------	----	6.0	190	335
	25°C, Ads.	---------	---	------	------	------	----	5.4	190	335
	45°C, Ads.	0.10-0.80	V	9.0473	0.5882	0.1058	4.47	3.5	190	335
	60°C, Ads.	---------	---	------	------	------	----	2.1	190	335
RAISINS	23.9°C, Ads.	0.10-0.74	III	0.9922	0.0414	0.0184	1.50	11.6	167	336

GULLE RAPESEED									
5°C, Ads.	0.30-0.90	VIII	5.3619	2.0696	0.0649	2.86	---	146	337
15°C, Ads.	0.30-0.90	VIII	5.2368	1.9624	0.0648	3.08	---	146	337
25°C, Ads.	0.30-0.90	VIII	5.2814	1.7141	0.0551	2.92	---	146	337
35°C, Ads.	0.40-0.90	VIII	5.4646	1.3261	0.0612	2.86	---	146	337
5°C, Des.	0.20-0.90	II	1.8205	20.3804	0.0264	1.83	---	146	338
15°C, Des.	0.20-0.90	VII	0.4619	6.2054	0.0428	2.78	---	146	338
25°C, Des.	0.20-0.90	II	1.7920	16.6671	0.0372	1.93	---	146	338
35°C, Des.	0.20-0.90	VII	0.4768	5.7488	0.0377	2.70	---	146	338
HEKTOR RAPESEED									
5°C, Ads.	0.40-0.90	II	1.7082	14.2157	0.0022	0.56	---	146	339
15°C, Ads.	0.40-0.90	II	1.7105	13.3906	0.0040	0.77	---	146	339
25°C, Ads.	0.40-0.90	II	1.7052	12.6126	0.0015	0.45	---	146	339
35°C, Ads.	0.40-0.90	VIII	5.3412	1.1916	0.0759	3.32	---	146	343
5°C, Des.	0.20-0.90	VII	0.4421	6.5844	0.0474	2.07	---	146	340
15°C, Des.	0.30-0.90	VIII	5.1521	2.3687	0.0445	1.75	---	146	341
25°C, Des.	0.30-0.90	VIII	5.2351	1.9643	0.0232	1.36	---	146	341
35°C, Des.	0.40-0.90	VIII	5.3097	1.6129	0.0232	1.58	---	146	341
TOWER RAPESEED									
5°C, Ads.	0.30-0.90	II	1.7702	19.4672	0.0081	0.63	---	146	342
15°C, Ads.	0.30-0.90	II	1.7269	16.8157	0.0050	0.58	---	146	342
25°C, Ads.	0.30-0.90	II	1.7187	15.6599	0.0029	0.67	---	146	342
35°C, Ads.	0.40-0.90	VIII	5.7509	1.4308	0.0764	2.60	---	146	343
5°C, Des.	0.30-0.90	II	1.9158	29.5925	0.0149	1.10	---	146	344
15°C, Des.	0.30-0.90	II	1.8594	24.8694	0.0401	1.71	---	146	344

Table I (Cont.)

PRODUCT	SPECIFICATIONS	RANGE OF a_w	Eq.	B(1)	B(2)	v	E% av	X_M % d.b.	REF.	FIG.
	25°C, Des.	0.30-0.90	II	1.8424	22.1267	0.0181	0.99	---	146	344
	35°C, Des.	0.40-0.90	II	1.7803	17.8831	0.0051	0.68	---	146	344
RASPBERRY	10°C, Ads. raw	0.10-0.80	IV	3.5698	1.1756	0.2695	5.56	2.8	50	345
RHUBARB	10°C, Ads.	0.10-0.80	VII	0.7160	9.1082	0.0146	1.90	5.1	50	346
RICE	25°C, (a)	0.10-0.80	III	2.7020	0.0006	0.1249	2.52	8.1	90	347
	25°C, (b)	0.10-0.80	VII	0.2973	11.8802	0.1184	3.02	7.3	90	347
	25°C, (a)	0.10-0.80	III	2.4936	0.0010	0.0101	0.79	7.9	188	348
	25°C, (b)	0.10-0.80	III	2.2202	0.0021	0.0241	1.35	8.1	33	348
	19.5°C, Des.	0.10-0.80	I	0.8632	6.0979	0.0405	1.31	8.1	174	349
	4.4°C, Ads.	0.10-0.80	I	0.8390	3.3439	0.0274	1.77	6.9	189	350
	4.4°C, Des.	-------	---	------	------	------	----	7.4	189	350
ROUGH RICE	25°C, (a)	0.10-0.80	III	2.1020	0.0040	0.0099	1.09	7.0	33	351
	25°C, (b)	0.20-0.80	III	2.5004	0.0012	0.0040	0.41	---	188	351
	26.7°C, Ads.	0.48-0.93	III	2.0244	0.0050	0.0115	0.56	---	62	352
	34.4°C, Ads.	0.52-0.93	III	1.9164	0.0071	0.0521	1.21	---	62	352
	43.9°C, Ads.	0.59-0.94	III	1.7080	0.0137	0.1899	2.37	---	62	352
	25°C, Ads.	0.40-0.90	VIII	4.7552	5.8483	0.0567	1.85	---	22	353
	0°C, Des.	0.20-0.90	I	0.8290	9.1228	0.0524	1.45	---	8	354
	20°C, Des.	0.20-0.90	III	2.4516	0.0013	0.0425	1.21	---	8	354
	30°C, Des.	0.20-0.90	III	2.3771	0.0018	0.0730	1.52	---	8	354

Condition	Range								Page
23.3°C, Des.	0.10-0.80	I	0.8389	6.0138	0.0564	2.01	7.1	59	355
25°C, Des.	0.15-0.90	III	2.0385	0.0046	0.0272	1.24	---	90	355
25°C, Des. (a)	0.10-0.80	III	2.1380	0.0034	0.0362	1.24	6.9	22	356
25°C, Des. (b)	0.15-0.90	I	0.8212	8.3094	0.0344	1.35	---	33	356
27.5°C, Ads.	0.44-0.84	VIII	4.8391	6.0576	0.0496	1.53	---	82	357
32.5°C, Ads.	0.43-0.84	VIII	4.9771	5.5973	0.0225	1.18	---	82	357
27.5°C, Des.	0.44-0.96	I	0.7862	13.6892	0.1821	2.05	---	82	358
32.5°C, Des.	0.43-0.96	I	0.7936	10.9897	0.1180	1.85	---	82	358
27.5°C, Des.	0.44-0.96	I	0.8045	9.7234	0.0577	1.20	---	82	359
32.5°C, Des.	0.43-0.96	I	0.8078	8.3425	0.0259	0.82	---	82	359
27.5°C, Ads.	0.44-0.84	VIII	4.9569	6.0403	0.0289	1.26	---	82	360
32.5°C, Ads.	0.43-0.84	VIII	5.2382	5.4600	0.0324	1.37	---	82	360
27.5°C, Ads.	0.44-0.84	VIII	5.2228	5.9443	0.0184	1.00	---	82	361
32.5°C, Ads.	0.43-0.84	VIII	5.4788	5.3723	0.0156	0.92	---	82	361
27.5°C, Des.	0.44-0.96	VIII	4.2949	8.1121	0.0789	1.57	---	82	362
32.5°C, Des.	0.43-0.96	I	0.8409	5.0592	0.0232	0.82	---	82	363
27.5°C, Ads.	0.44-0.84	VIII	5.4461	5.6497	0.0038	0.40	---	82	364
32.5°C, Ads.	0.43-0.84	VIII	5.6015	5.2005	0.0107	0.77	---	82	364
27.5°C, Des.	0.44-0.96	I	0.8280	6.6846	0.0559	1.21	---	82	365
32.5°C, Des.	0.43-0.96	I	0.8346	5.6226	0.1331	1.80	---	82	365

BURMESE RICE BRAN

Condition	Range								Page
15°C, Ads.	0.45-0.90	VIII	5.4239	5.6106	0.0398	1.23	---	144	366
25°C, Ads.	0.48-0.90	VIII	5.5300	5.0556	0.0541	1.51	---	144	366
35°C, Ads.	0.50-0.90	VIII	5.4316	4.9435	0.0531	1.44	---	144	366

Table I (Cont.)

PRODUCT	SPECIFICATIONS	RANGE OF a_w	Eq.	B(1)	B(2)	v	E% av	X_M % d.b.	REF.	FIG.
	15°C, Des.	0.34-0.89	VII	0.2728	12.1416	0.0863	1.66	---	144	367
	25°C, Des.	0.37-0.90	VII	0.2827	11.6724	0.0745	1.43	---	144	367
	35°C, Des.	0.40-0.90	VIII	4.6777	6.8276	0.0638	1.65	---	144	366
INDIAN RICE										
BRAN	15°C, Ads.	0.42-0.89	II	2.1302	80.4329	0.1217	2.21	---	144	368
	25°C, Ads.	0.45-0.90	II	2.1522	78.5154	0.0743	1.69	---	144	368
	35°C, Ads.	0.47-0.91	II	2.1971	81.3531	0.0434	1.28	---	144	368
	15°C, Des.	0.32-0.89	VII	0.3165	11.2401	0.0792	1.60	---	144	368
	25°C, Des.	0.35-0.90	VIII	5.2040	6.0264	0.0717	1.85	---	144	369
	35°C, Des.	0.39-0.91	VIII	5.0795	5.8459	0.0884	2.15	---	144	369
SOUTH AFRICAN										
RICE BRAN	15°C, Ads.	0.39-0.92	VIII	4.8202	4.4943	0.0600	1.63	---	144	371
	25°C, Ads.	0.42-0.92	II	2.2986	85.0347	0.0358	1.32	---	144	370
	35°C, Ads.	0.45-0.92	II	2.2045	64.3119	0.0423	1.52	---	144	370
	15°C, Des.	0.36-0.92	VIII	4.6980	4.8542	0.0342	1.06	---	144	371
	25°C, Des.	0.39-0.92	VIII	4.8202	4.4943	0.0600	1.63	---	144	371
	35°C, Des.	0.41-0.92	VIII	4.9515	4.1645	0.0651	1.81	---	144	371
TANZANIAN RICE										
BRAN	15°C, Ads.	0.55-0.89	VI	-1.4700	6.2673	0.1827	2.43	---	144	372
	25°C, Ads.	0.59-0.90	VI	-1.4188	5.9644	0.1296	2.12	---	144	372
	35°C, Ads.	0.62-0.90	V	1.4821	6.0655	0.1820	2.58	---	144	372

15°C, Des.	0.38-0.83	VIII	6.2034	4.8983	0.0105	0.72	---	144	373
25°C, Des.	0.41-0.84	VIII	6.1495	4.5832	0.0162	0.88	---	144	373
35°C, Des.	0.43-0.85	VIII	6.0082	4.4733	0.0169	0.85	---	144	373
SACC. CEREVISIAE 20°C, Ads.(a)	0.10-0.67	VI	-5.3564	1.4603	0.0552	2.64	4.4	135	374
20°C, Ads. (b)	0.10-0.67	VI	-5.0269	0.8324	0.0228	1.82	3.8	135	374
20°C, Des.	0.10-0.67	V	5.2183	2.4657	0.0328	2.56	3.5	135	374
SAFFLOWER PROTEIN 35°C, Ads.=Des	0.10-0.80	III	1.9405	0.0089	0.0782	3.80	6.5	77	375
SAFFLOWER PROTEIN + "AVICEL" 35°C, Ads.	0.10-0.80	I	0.7787	4.5915	0.0351	3.23	5.1	77	376
SAFFLOWER PROTEIN + STARCH 35°C, Ads.	--------	---	------	------	------	----	7.6	77	377
SAFFLOWER PROTEIN + STARCH GEL 35°C, Ads.	0.10-0.80	III	1.7490	0.0111	0.0143	0.93	6.0	77	378
35°C, Des.	0.10-0.80	III	2.0792	0.0038	0.0279	1.05	6.9	77	378
SALMIN 25°C, Ads.	--------	---	------	------	------	----	5.9	24	379
40°C, Ads.	--------	---	------	------	------	----	5.5	24	379
SALMON 37°C, Ads.	--------	---	------	------	------	----	6.1	120	380

Table I (Cont.)

PRODUCT	SPECIFICATIONS	RANGE OF a_w	Eq.	B(1)	B(2)	v	E% av	X_M % d.b.	REF.	FIG.
SALSIFY	45°C, Ads.	0.10-0.80	II	1.3293	15.1924	0.0184	1.18	5.2	190	381
	60°C, Ads.	0.10-0.80	II	1.0917	6.8883	0.0810	2.51	4.2	190	381
	45°C, Des.	0.10-0.80	II	1.3738	18.8155	0.0143	0.58	5.7	190	382
BEEF SARCOPLAS-MIC FRACTION	11.1°C, Des.	----------	---	------	------	------	----	15.2	133	383
	21.1°C, Des.	----------	---	------	------	------	----	12.3	133	383
SORGHUM	4.4°C, Ads.	0.10-0.80	I	0.8450	7.4799	0.0091	0.59	8.1	26	384
	21.1°C, Ads.	0.10-0.80	I	0.8336	7.4660	0.0020	0.28	7.5	26	384
	32.2°C, Ads.	0.10-0.80	III	2.3218	0.0022	0.0033	0.45	7.0	26	384
	25°C, Ads.	0.22-0.85	IV	1.2862	2.5860	0.0307	1.09	---	6	385
	35°C, Ads.	0.25-0.86	I	0.8219	7.3452	0.0296	1.05	---	6	385
	25°C, Des.	0.18-0.88	III	2.0596	0.0034	0.0874	1.51	---	6	386
	35°C, Des.	0.20-0.88	III	1.9558	0.0048	0.0586	1.14	---	6	386
	25°C, Ads.	0.27-0.81	VII	0.3030	11.7448	0.0076	0.52	---	6	387
	35°C, Ads.	0.29-0.82	VII	0.3049	11.3138	0.0070	0.49	---	6	387
	25°C, Des.	0.25-0.86	III	1.9968	0.0045	0.0170	0.58	---	6	388
	35°C, Des.	0.28-0.87	III	1.9092	0.0061	0.0125	0.56	---	6	388
COMPOUNDED SOUP	20°C, Sorp.	----------	---	------	------	------	----	5.1	50	389
MEAT AND VEGETA-BLE SOUP	37°C, Sorp.	0.10-0.80	VII	0.7883	9.7198	0.0865	3.66	---	50	391

	Conditions	Range	Eq.							Page
VEGETABLE SOUP	37°C, Sorp.	0.10-0.80	IV	3.4000	1.4131	0.0359	1.95	3.8	50	392
SOYBEANS	30°C, Ads.	0.10-0.80	VII	0.6857	5.1471	0.0585	5.89	2.6	165	393
	30°C, Des.	0.10-0.85	VII	0.5637	5.9831	0.1864	4.93	3.2	165	393
	30°C, Ads.	0.10-0.80	II	1.2603	8.8080	0.0920	2.53	4.1	165	394
	30°C, Des.	0.10-0.85	II	1.3425	11.3470	0.0598	2.01	4.2	165	394
	30°C, Ads.	0.10-0.80	VII	0.5548	6.0027	0.1177	3.80	3.2	165	395
	30°C, Des.	0.10-0.80	VII	0.5304	6.5204	0.0953	5.02	3.5	165	395
	25°C, Ads.=Des	0.25-0.80	II	1.5385	18.1424	0.0052	0.74	---	140	396
SOYA MEAL	15°C, Ads.	0.30-0.86	VI	-4.0861	4.0649	0.0147	0.67	---	143	397
	35°C, Ads.	0.36-0.87	VI	-3.7916	3.9532	0.1089	2.14	---	143	397
SOYBEAN PRODUCTS	25°C, Des. (a)	0.30-0.90	II	1.3805	16.2320	0.1300	2.64	---	143	399
	25°C, Des. (b)	0.40-0.80	II	1.5051	17.3633	0.0061	0.57	---	143	399
	25°C, Ads. (a)	0.40-0.90	II	1.3269	13.4399	0.0829	1.82	---	143	400
	25°C, Ads. (b)	0.40-0.80	II	1.5286	17.4814	0.0031	0.45	---	143	400
SOYBEAN SEED	15°C, Sorp.	0.10-0.80	II	1.5817	20.9621	0.0842	4.04	5.0	49	401
SOY PROTEIN CONCENTRATE	1°C, Ads.	0.07-0.80	VII	0.3791	12.9405	0.0961	2.58	7.0	54	402
	21°C, Ads.	0.08-0.81	VII	0.4437	11.5443	0.1491	3.16	6.7	54	402
	37°C, Ads.	0.09-0.80	VII	0.4870	10.6769	0.1729	3.49	5.7	54	402

Table I (Cont.)

PRODUCT	SPECIFICATIONS	RANGE OF a_w	Eq.	B(1)	B(2)	v	E% av	X_M % d.b.	REF.	FIG.
SPINACH	10°C, Ads.	0.10-0.80	II	1.4265	20.5440	0.1221	1.95	5.5	50	406
	37°C, Ads.=Des	0.10-0.70	VI	-6.0909	0.6823	0.0159	1.40	4.5	119	407
STARCH	30°C	0.10-0.84	III	2.3241	0.0016	0.1756	3.41	9.4	115	408
	80°C	0.10-0.84	III	1.9557	0.0070	0.1419	4.35	7.4	115	408
	35°C, Ads.	0.10-0.80	I	0.8545	4.0844	0.0160	1.10	6.9	77	409
STARCH GEL	-20 = -10 =									
	= 0°C, Ads.	0.10-0.80	III	2.9305	0.0001	0.2718	3.00	11.4	166	410
	10°C, Ads.	0.10-0.80	III	2.3731	0.0009	0.0672	1.81	10.1	166	410
	20°C, Ads.	0.10-0.80	III	2.2011	0.0018	0.0864	2.14	9.2	166	410
	30°C, Ads.	---------	---	------	------	------	----	8.2	166	410
	40°C, Ads.	0.10-0.80	III	2.4839	0.0015	0.1591	3.63	7.8	166	410
	50°C, Ads.	---------	---	------	------	------	----	7.2	166	410
	25°C	0.10-0.80	III	2.0178	0.0040	0.0679	2.58	8.7	45	411
	35°C, Ads.	0.10-0.80	III	1.8372	0.0069	0.0231	1.25	7.1	77	412
	35°C, Des.	0.10-0.80	III	1.9807	0.0032	0.0469	1.46	8.8	77	412
STARCH GEL + "AVICEL"	35°C, Ads.	0.10-0.80	III	1.7717	0.0148	0.0134	1.51	4.8	77	413
STARCH-GLUCOSE GEL	20°C, Ads.	0.20-0.80	IV	2.9508	1.4390	0.0421	1.47	---	166	414
	30°C, Ads.	0.20-0.80	IV	2.8693	1.6143	0.0069	0.64	---	166	414

	Condition	Range								
	40°C, Ads.	0.20-0.80	III	1.0423	0.1050	0.0206	1.74	---	166	414
	50°C, Ads.	0.20-0.80	III	1.1074	0.1744	0.0169	2.36	---	166	414
MAIZE STARCH	25°C, Ads.	0.34-0.87	VIII	5.4823	8.8112	0.1158	1.72	---	170	415
	30°C, Ads.	0.33-0.92	I	0.8275	8.6705	0.1070	1.48	---	170	416
	50°C, Ads.	0.32-0.87	I	0.8388	5.8369	0.0219	0.71	---	170	416
POTATO STARCH	25°C, Ads.	0.34-0.87	VIII	7.9664	10.2453	0.0618	0.98	---	170	417
	40°C, Ads.	0.32-0.89	III	2.5286	0.0006	0.0439	0.95	---	170	417
	25°C, Ads.(a)	0.10-0.80	III	1.2844	0.0415	0.1086	3.21	6.0	28	419
	25°C, Ads.(b)	0.10-0.80	VII	0.4968	10.4963	0.0668	3.26	6.9	28	419
	25°C, Ads.(c)	0.09-0.81	I	0.8668	3.2396	0.0621	2.19	7.2	28	419
WHEAT STARCH	25°C, Ads.	0.34-0.87	I	0.7886	14.2468	0.1019	1.55	---	170	421
	40°C, Ads.	0.32-0.89	III	2.4265	0.0015	0.0745	1.46	---	170	421
	50°C, Ads.	0.32-0.87	I	0.8363	5.9465	0.0086	0.52	---	170	420
	20.2°C, Ads.	0.10-0.92	VII	0.3789	12.4154	0.0189	0.78	7.0	25	422
	30.1°C, Ads.	0.10-0.92	VII	0.3850	11.8319	0.0053	0.54	6.7	25	422
	40.8°C, Ads.	0.11-0.92	III	1.9396	0.0061	0.1554	1.76	6.6	25	422
	50.2°C, Ads.	0.12-0.91	III	1.8544	0.0083	0.0244	0.97	6.6	25	422
	27°C, Ads.	0.10-0.80	I	0.8468	5.7637	0.0244	1.31	7.1	25	423
	27°C, Des.	0.10-0.80	III	2.4923	0.0009	0.0181	0.98	8.2	25	423
STRAWBERRY	25°C, Ads.	0.10-0.75	IV	3.3110	1.7357	0.0757	2.47	7.7	107	424

Table I (Cont.)

PRODUCT	SPECIFICATIONS	RANGE OF a_w	Eq.	B(1)	B(2)	v	E% av	X_M % d.b.	REF.	FIG.
SUCROSE	10°C, Ads.	0.10-0.80	VII	0.8481	12.0051	0.0688	1.97	8.0	50	425
	20°C, Ads.	0.47-0.85	V	0.0359	-0.0183	0.0004	30.14	---	111	426
	80°C	--------	--	-----	-----	-----	-----	1.0	115	427
	35°C, Ads.	0.10-0.80	III	0.8749	0.0676	0.2287	4.11	9.9	66	428
	47°C, Ads.	--------	--	-----	-----	-----	-----	7.8	66	428
	25°C, Ads.	0.10-0.80	VII	0.8284	10.7708	0.0255	1.41	7.3	172	429
CANE SUGAR PRODUCTS	25°C, Sorp.(a)	--------	--	----	----	----	----	0.8	152	430
	25°C, Sorp.(b)	--------	--	----	----	----	----	0.5	152	430
	25°C, Sorp.(c)	--------	--	----	----	----	----	0.2	152	430
ICING SUGAR	15°C, Ads.	--------	--	----	----	----	----	6.4	158	431
	10°C, Ads.	0.10-0.80	II	1.0095	3.9626	0.0142	1.64	3.1	158	432
	20°C, Ads.	0.10-0.80	II	1.0258	4.3225	0.2686	5.83	3.4	158	433
	20°C, Des.	--------	--	----	----	----	----	5.6	158	433
	20°C, Ads.	0.10-0.80	V	1.8703	1.2313	0.0148	2.87	1.6	158	435
	20°C, Des.	0.10-0.80	V	1.7213	2.3084	0.0370	3.91	---	158	435
	20°C, Ads.	0.10-0.80	V	1.9079	0.6863	0.0178	4.99	1.2	158	436
	20°C, Des.	0.10-0.80	V	1.8865	1.7263	0.0326	4.07	1.9	158	436
SULTANAS	5 = 35°C, Ads.	0.41-0.72	II	0.8127	5.4512	0.0669	1.58	---	141	437

SUNFLOWER SEED	25°C, Ads.	0.25-0.85	II	1.7036	14.2094	0.0288	2.56	---	140	439
SWEET MARJORAM	5°C, Ads.	0.10-0.80	II	1.9237	68.8515	0.0940	2.54	6.3	190	440
	25°C, Ads.	0.10-0.80	II	1.4701	17.2126	0.0756	2.22	4.6	190	440
	45°C, Ads.	0.10-0.80	II	1.1235	5.5094	0.1529	3.64	3.0	190	440
	60°C, Ads.	-------	---	------	-------	------	----	2.1	190	440
	5°C, Des.	0.10-0.80	II	2.3930	333.8907	0.0686	1.74	7.4	190	441
	25°C, Des.	0.10-0.80	II	1.5945	26.1584	0.0803	1.73	5.2	190	441
	45°C, Des.	0.10-0.80	II	1.1235	5.5094	0.1529	3.64	3.1	190	441
TAPIOCA	45°C, Ads.	0.10-0.80	III	1.3997	0.0317	0.0590	4.14	6.4	190	442
	25°C, Des.	-------	---	------	-------	------	----	8.7	190	443
	45°C, Des.	0.10-0.80	III	1.6565	0.0143	0.1204	4.17	6.9	190	443
BLACK TEA	21°C, Ads.	0.10-0.80	II	1.2831	7.1343	0.1336	4.23	3.1	79	444
	21°C, Des.	0.10-0.80	VI	-2.6706	4.8585	0.0828	2.59	6.5	79	444
	32°C, Ads.	0.10-0.80	II	1.3851	9.6222	0.0773	3.45	3.6	79	445
	32°C, Des.	0.10-0.80	II	2.0195	62.2539	0.0526	2.54	5.6	79	445
	21°C, Ads.	0.10-0.80	VI	-3.2246	0.6756	0.0960	5.71	2.4	79	446
	21°C, Des.	0.10-0.80	VI	-2.6817	4.6521	0.0758	2.75	6.5	79	446
	32°C, Ads.	0.10-0.80	II	1.4985	13.4015	0.0871	2.85	3.7	79	447
	32°C, Des.	0.10-0.80	II	2.2622	115.2066	0.1477	3.84	6.3	79	447
TEA FANNINGS	15°C, Sorp.	---------	---	------	-------	------	----	5.2	50	448

Table I (Cont.)

PRODUCT	SPECIFICATIONS	RANGE OF a_w	Eq.	B(1)	B(2)	v	E% av	X_M % d.b.	REF.	FIG.
TEA INFUSION	37°C, Sorp.	---------	---	------	------	------	----	4.9	50	448
	10°C, Ads. (a)	0.10-0.80	VII	0.7786	8.5960	0.0550	2.57	5.9	50	449
	10°C, Ads. (b)	0.10-0.80	II	1.1094	7.8124	0.0670	1.76	4.4	50	449
TEA, leaves	20°C	---------	---	------	------	------	----	3.6	80	450
THYME	5°C, Ads.	0.10-0.80	VII	0.4043	10.2073	0.0297	1.85	5.6	190	451
	25°C, Ads.	0.10-0.80	II	1.5840	21.5361	0.0447	2.65	4.7	190	451
	45°C, Ads.	0.10-0.80	II	1.3091	9.4088	0.0039	1.02	3.5	190	451
	60°C, Ads.	0.10-0.80	V	3.6166	2.3982	0.1271	4.07	3.1	190	451
	5°C, Des.	0.10-0.80	VII	0.3077	12.1737	0.0366	1.25	7.0	190	452
	25°C, Des.	0.10-0.80	II	1.7539	34.9373	0.0148	1.26	4.9	190	452
	45°C, Des.	0.10-0.80	II	1.3091	9.4088	0.0039	1.02	3.6	190	452
TOMATO	17°C, Ads.	0.10-0.80	II	0.9704	10.0587	0.1099	3.02	8.2	3	453
	27°C, Ads.	---------	---	------	------	------	----	6.1	3	453
	10°C, Ads.	---------	---	------	------	------	----	3.2	50	454
TROUT, cooked	5°C, Ads.	0.10-0.80	I	0.8842	3.4626	0.0882	2.16	8.6	190	455
	45°C, Ads.	0.10-0.80	VII	0.5691	8.2126	0.0718	3.85	4.3	190	455
	60°C, Ads.	0.10-0.80	VI	-3.7359	1.0946	0.0200	1.76	3.3	190	455
	45°C, Des.	0.10-0.80	II	1.5526	20.3488	0.0422	2.50	4.4	190	456
TROUT, raw	5°C, Ads.	0.10-0.80	III	1.3857	0.0176	0.0626	1.81	7.9	190	457

Item	Condition	Range								
	45°C, Ads.	0.10-0.80	II	1.1329	8.0277	0.1000	4.36	4.3	190	457
	60°C, Ads.	0.10-0.80	V	5.3728	1.6769	0.0630	3.25	3.5	190	457
	5°C, Des.	0.10-0.80	VII	0.3871	15.3339	0.1208	2.00	8.8	190	458
	45°C, Des.	0.10-0.80	VII	0.3903	15.2898	0.1370	2.18	8.8	190	458
	60°C, Des.	0.10-0.80	V	5.3728	1.6769	0.0630	3.25	3.5	190	458
BIG-EYE TUNA	25°C, Ads.	0.10-0.80	II	1.1702	11.3017	0.0231	1.34	5.4	93	459
TURKEY, cooked	22°C, Ads.	0.10-0.80	VII	0.4948	8.8714	0.0208	2.18	5.6	92	460
	22°C, Des.	0.10-0.80	VII	0.4864	10.0165	0.0638	2.94	6.5	92	460
	10°C, Des.	0.10-0.80	III	1.4720	0.0160	0.0257	1.66	7.8	92	461
	0°C, Des.	0.10-0.80	III	1.3296	0.0201	0.0623	2.32	8.8	92	461
WAFER SHEET	24°C, Ads.	0.10-0.75	III	1.8280	0.0155	0.0593	2.96	5.0	10	462
WALNUT KERNELS	25°C, Sorp.	0.11-0.81	II	2.1303	9.6452	0.0255	3.42	1.9	176	463
	25°C, Sorp.	0.11-0.81	VII	0.3959	3.1729	0.0035	2.00	1.8	176	464
	25°C, Sorp.	0.11-0.81	I	0.5644	4.8961	0.0110	3.33	1.6	176	465
IN-SHELL WALNUTS	22.5°C, Ads.	0.10-0.77	VII	0.4920	6.9754	0.1323	6.19	4.9	155	466
	22.5°C, Des.	0.10-0.80	I	0.8104	3.7591	0.0520	2.59	5.8	155	466
SHELLED WALNUTS	7°C, Ads.	0.10-0.80	II	2.0981	10.2130	0.0260	4.61	2.3	155	467
	22.5°C, Ads.	0.10-0.80	II	2.7222	20.0611	0.0093	2.76	2.1	155	467
WHEAT	35°C, Ads.	0.10-0.80	III	1.8349	0.0092	0.0047	0.77	6.0	63	468
	25°C, Des.	0.10-0.80	III	2.3861	0.0014	0.0132	0.85	7.9	63	468

Table I (Cont.)

PRODUCT	SPECIFICATIONS	RANGE OF a_w	Eq.	B(1)	B(2)	v	E% av	X_M % d.b.	REF.	FIG.
	35°C, Des.	0.10-0.80	III	2.3636	0.0019	0.0063	0.54	7.1	63	468
	30°C, Ads.	0.10-0.80	III	1.9145	0.0064	0.0089	0.97	7.0	63	469
	30°C, Des.	0.10-0.80	III	2.3611	0.0015	0.0287	1.38	8.3	63	469
	25°C, Des.	0.09-0.86	VII	0.3205	13.0959	0.0732	2.14	7.8	13	470
	50°C, Des.	0.13-0.82	VII	0.3906	10.4073	0.0599	2.18	6.0	13	470
	18.3°C, Ads. (a)	0.10-0.80	VII	0.3109	11.7525	0.0596	1.59	6.7	44	471
	18.3°C, Ads. (b)	0.10-0.80	VII	0.3043	11.6542	0.0809	2.07	6.6	44	471
	18.3°, Ads. (c)	0.10-0.80	II	2.1817	125.3284	0.0302	1.52	6.2	44	471
	18.3°C, Ads. (d)	0.10-0.80	II	2.1182	100.3737	0.0245	1.43	5.9	44	471
	25.5°C, Ads. (a)	0.10-0.80	VII	0.3204	10.9021	0.0438	1.57	6.4	44	472
	25.5°C, Ads. (b)	0.10-0.80	VII	0.3064	11.0005	0.0516	1.72	6.4	44	472
	25.5°C, Ads. (c)	0.10-0.80	VII	0.3354	10.5781	0.1126	2.75	6.1	44	472
	25.5°C, Ads. (d)	0.10-0.80	VII	0.3383	10.5908	0.1291	2.98	6.1	44	472
	32.2°C, Ads. (a)	0.10-0.80	VII	0.3186	10.8410	0.0716	2.19	6.5	44	473

32.2°C, Ads. (b)	0.10-0.80	VII	0.3239	10.6275	0.1097	2.57	6.4	44	473
32.2°C, Ads. (c)	0.10-0.80	II	2.0802	82.7396	0.1212	3.50	6.0	44	473
32.2°C, Ads. (d)	0.10-0.80	VII	0.3439	10.3841	0.1305	3.01	6.0	44	473
25°C, Ads.	0.25-0.85	I	0.8551	5.1552	0.0091	0.54	---	139	474
25°C, Ads.	0.49-0.93	VIII	5.8767	7.4543	0.0351	0.97	---	6	475
35°C, Ads.	0.52-0.86	VIII	5.7031	7.2701	0.0062	0.21	---	6	475
25°C, Des.	0.43-0.91	VIII	5.7185	8.0577	0.0304	1.04	---	6	475
25°C, Des.	0.44-0.91	VIII	5.9279	7.1083	0.0455	0.98	---	6	476
35°C, Ads.	0.53-0.93	II	2.4558	289.5902	0.0708	0.87	---	6	477
WHEAT FEED									
15°C, Ads.	0.40-0.90	VIII	7.3702	4.3746	0.0562	1.67	---	185	479
15°C, Des.	0.40-0.90	VIII	6.3591	6.7906	0.0145	0.76	---	185	479
35°C, Des.	0.40-0.90	VIII	6.8799	5.4708	0.0203	0.72	---	185	479
35°C, Ads.	0.40-0.90	II	1.8066	41.7388	0.0534	1.48	---	185	480
WHEAT FLOUR									
27°C, Ads.	0.10-0.80	VII	0.3068	11.8805	0.0060	0.60	6.7	25	481
27°C, Des.	0.10-0.80	I	0.8170	10.2970	0.0137	0.87	7.9	25	481
20.2°C, Ads.	0.12-0.89	II	2.3122	200.5532	0.0396	1.66	5.9	25	482
30.1°C, Ads.	0.13-0.90	VII	0.3441	11.4040	0.1096	2.32	5.9	25	482
40.8°C, Ads.	0.13-0.90	VII	0.3605	10.8568	0.0728	2.12	5.6	25	482
50.2°C, Ads.	0.15-0.90	VII	0.3908	10.0209	0.1042	1.95	5.2	25	482

Table I (Cont.)

PRODUCT	SPECIFICATIONS	RANGE OF a_w	Eq.	B(1)	B(2)	v	E% av	X_M % d.b.	REF.	FIG.
WHEY POWDERS	24.5°C, Ads.	-------	---	------	------	------	----			
	(a)	--------	---	------	------	------	----	2.4	14	484
	24.5°C, Ads.	--------	---	------	------	------	----			
	(b)	--------	---	------	------	------	----	2.4	14	484
WHEY PROTEIN CONCENTRATE	24°C, Ads. (a)	0.12-0.86	II	1.4806	17.6165	0.2190	4.23	4.8	17	487
	24°C, Ads. (b)	0.09-0.86	VII	0.4563	11.9073	0.1588	2.45	6.8	17	487
	24°C, Ads. (a)	0.10-0.80	II	1.0367	8.4317	0.1632	4.36	5.0	17	488
	24°C, Ads. (b)	0.10-0.80	II	1.3573	15.9825	0.0147	1.43	5.0	17	488
	24°C, Ads. (c)	0.10-0.80	II	1.3893	17.3629	0.0536	1.90	5.0	17	488
SWEET WHEY	24.5°C, Ads.	--------	---	------	------	------	----	8.1	14	489
	24.5°C, Des.	0.10-0.80	V	3.1279	3.0619	0.0601	3.73	---	14	489
SWEET WHEY-SOY BEVERAGE	24°C, Sorp.	--------	---	------	------	------	----	5.8	18	490
WINTER SAVORY	5°C, Ads.	0.10-0.80	VII	0.3282	12.7681	0.0393	1.76	7.3	190	491
	25°C, Ads.	0.10-0.80	II	1.9929	89.5484	0.1144	2.74	6.5	190	491
	45°C, Ads.	--------	---	------	------	------	----	3.7	190	491
	60°C, Ads.	0.10-0.80	V	2.8755	2.2015	0.1476	5.15	2.9	190	491
	5°C, Des.	0.10-0.80	VII	0.2275	15.0949	0.0185	0.87	9.1	190	492
	25°C, Des.	0.10-0.80	II	2.3329	252.6355	0.0194	0.85	7.0	190	492
	45°C, Des.	--------	---	------	------	------	----	3.5	190	492

YAMS	37°C,	Ads.=Des	0.10-0.70	II	1.2423	10.6590	0.0168	1.60	4.8	119	493
ACTIVE DRIED YEAST											
	16°C,	Ads.	--------	--	-----	------	------	---	5.9	135	494
	27°C,	Ads.	--------	---	-----	------	------	---	5.6	135	494
	44°C,	Ads.	--------	---	-----	------	------	---	5.5	135	494
YEAST											
	16°C,	Des.	0.10-0.75	V	6.3072	4.9435	0.1047	2.32	5.6	135	495
	27°C,	Des.	0.10-0.75	V	6.0234	4.1318	0.1813	3.30	4.8	135	495
	44°C,	Des.	0.10-0.75	VI	-5.3953	1.5535	0.0570	3.13	5.0	135	495
	20°C,	Ads.	0.07-0.75	V	5.6294	2.8108	0.0574	3.53	3.7	111	496
	10°C,	Ads.	0.10-0.80	II	1.3126	16.2487	0.0360	2.04	6.2	50	497
TROPINA YEAST											
	10°C,	Ads.	0.31-0.85	II	1.3553	15.1013	0.0701	1.68	---	147	498
	25°C,	Ads.	0.34-0.85	II	1.3018	12.5683	0.0518	1.34	---	147	498
	40°C,	Ads.	0.35-0.85	II	1.2758	11.4324	0.0871	1.78	---	147	498
	10°C,	Des.	0.26-0.85	VII	0.5557	10.7330	0.0648	1.91	---	147	499
	25°C,	Des.	0.30-0.85	II	1.3873	16.7734	0.0805	1.78	---	147	499
	40°C,	Des.	0.32-0.85	II	1.3279	13.6909	0.0770	1.56	---	147	499
YOGHURT											
	5°C,	Ads.	0.10-0.80	IV	2.8402	1.6808	0.0744	2.24	5.2	190	500
	25°C,	Ads.	0.10-0.80	II	1.0529	6.4806	0.1203	2.63	4.1	190	500
	45°C,	Ads.	0.10-0.80	VI	-3.6752	0.2732	0.0229	3.84	3.0	190	500
	5°C,	Des.	0.10-0.80	IV	2.5442	1.9048	0.0980	2.71	5.4	190	501
	25°C,	Des.	0.10-0.80	II	1.0529	6.4806	0.1203	2.63	4.2	190	501

Letters within brackets correspond to those shown in the figures.

Table II. Complimentary Table for Iglesias and Chirife's (1978) Eq. (IV)

PRODUCT	TEMP.	RANGE OF a_w	$X_{0.50}$	REF.	FIG.
BANANA	10°C, Ads. scalded	0.10-0.80	10.90	50	35
CABBAGE	19.5°C, Des.	0.10-0.79	14.00	174	60
CARROTS	37°C, Ads. = Des.(b)	0.10-0.70	11.90	119	68
	70°C, Ads. = Des.(c)	0.10-0.60	10.40	119	68
	37°C, Sorp.	0.10-0.70	10.80	50	70
	60°C, Sorp.	0.10-0.80	9.10	50	70
	22°C, Ads.	0.09-0.80	10.10	112	74
CITRUS JUICE	25°C, Ads. (d)	0.11-0.75	10.53	94	112
MALTOSE	23°C, Ads.	0.10-0.76	11.35	46	226
PEAR	25°C, Ads.	0.10-0.80	15.00	190	297
	25°C, Des.	0.10-0.80	15.00	190	297
	25°C, Des.	0.20-0.80	14.80	190	298
PINEAPPLE	60°C, Ads.	0.10-0.80	10.90	190	309

Table II (Cont.)

PRODUCT	TEMP.	RANGE OF a_w	$X_{0.50}$	REF.	FIG.
RASPBERRY	10°C, Ads.	0.10-0.80	9.00	50	345
SORGHUM	25°C, Ads.	0.22-0.85	12.17	6	385
VEGETABLE SOUP	37°C, Sorp.	0.10-0.80	10.90	50	392
STARCH-GLUCOSE GEL	20°C, Ads.	0.20-0.80	9.19	166	414
	30°C, Ads.	0.20-0.80	10.43	166	414
STRAWBERRY	25°C, Ads.	0.10-0.75	14.54	107	424
YOGHURT	5°C, Ads.	0.10-0.80	11.30	190	500
	5°C, Des.	0.10-0.80	11.70	190	501

Letters within brackets correspond to those shown in the figures.

III. APPENDIX: NONLINEAL REGRESSION PROGRAM USED FOR DETERMINATION OF PARAMETERS B(1) AND B(2)

The two variables used are X, which stands for a_w (water activity) independent variable, and Y used for moisture content, percent dry or wet basis, the dependent variable.

In the subroutine F code, F is the function explicitly given for the dependent variable. As an example, Smith's equation is used. Consequently,

$$X = B(2) - B(1) \ln(1 - a_w). \tag{VIII}$$

Therefore,

$$F = B(2) - B(1) * A \log\{1 - [X(I, 1)]\}.$$

In the subroutine P code, two derivatives are included:

$$P(1) = \frac{\partial F}{\partial [B(1)]},$$

$$P(2) = \frac{\partial F}{\partial [B(2)]}.$$

Therefore for the above-mentioned example,

$$P(1) = -A \log\{1 - [X(I, 1)]\},$$
$$P(2) = 1.$$

The data cards, included after the program, should be as follows:

Card columns

Card 1 :   1–3, total number of data
           4–6, number of parameters
           7–9, number of constants
           10–12, independent variables
           13–15 = 0
Card 2 :   Blank
Card 3 :   Blank
Card 4 :   Initial values of parameters
Card 5 :   (8F 10.0)
Card 6 :   Water activity values
Card 7 :   Moisture content values

Remark: Although throughout the text X was used to denote moisture content values, as an exception it stands for a_w only in the computer program nomenclature.

```
// OPTION LIST,LINK
 INCLUDE ILFGHTAB
// EXEC FFORTRAN
      DIMENSION X(100,10),Y(100)
      DIMENSION BS(20),DB(20),BA(20),G(20),W(21),IB(19),SA(20),
     1P(20),A(20,21),B(20)
      DIMENSION FMT(18),PRNT(5),SPRNT(5)
 6500 XLL=0.
      GAMMA=0.
      NPRNT=0
  650 IWHER=0
  652 GO TO 4
  653 IWHER=IWHER
      IF(IWHER.GT.0) GO TO 654
      IF(IWHER.EQ.0) GO TO 660
  651 CONTINUE
      CALL SUBZ(Y,X,B,PRNT,NPRNT,N)
      IF(IBOUT.EQ.0) GO TO 652
      GO TO 650
  654 CONTINUE
      CALL FCODE(Y,X,B,PRNT,F,I)
 3000 IF(IWHER.NE.1) GO TO 652
  656 IF(IFSS2.NE.0) GO TO 652
  658 CONTINUE
      CALL PCODE(P,X,B,PRNT,F,I)
      GO TO 652
  660 CALL EXIT
    4 IWHER=IWHER
      IF(IWHER.LT.0) GO TO 59
      IF(IWHER.EQ.0) GO TO 10
    8 GO TO ( 75,304,606,620), IWHER
   10 ITOT=0
      IBOUT=0
      READ(5,900,END=660)N,K,IP,M,IFP
      IF(N.LE.0) GO TO 20
      READ(5,900)IWS1,IWS2,IWS3,IWS4,IWS5,IWS6
      IFSS1=2
      IF(IWS5.EQ.0)GO TO 210
  210 CONTINUE
      WRITE (6,932)
      IF(IFSS1.NE.1) GO TO 211
      WRITE(6,932)
  211 GO TO 21
   20 GO TO (19,17),IBKT
   17 CONTINUE
   19 IWHER=0
      GO TO 653
   21 IF(IFP.LE.0) GO TO 22
   23 CONTINUE
   22 IF(IP.LE.0) GO TO 30
   24 READ(5,900)(IB(I),I=1,IP)
      DO 26 I=1,IP
      IF(IB(I).GT.0) GO TO 26
   25 WRITE (6,926)
```

```
      IF(IFSS1.NE.1) GO TO 212
      WRITE (6,926)
  212 CONTINUE
      IBOUT=1
   26 CONTINUE
   30 READ(5,931)FF,T,E,TAU,XL,GAMCR,DEL,ZETA
      IF(FF.GT.0.) GO TO 34
   32 FF=4
   34 IF(E.GT.0.) GO TO 37
   36 E=0.00005
   37 IF(TAU.GT.0.) GO TO 39
   38 TAU=.001
   39 IF(T.GT.0.) GO TO 42
   40 T=2.
   42 IF(K.GT.25) GO TO 46
   44 IBKT=1
      GO TO 50
   46 IBKT=2
   50 IF(GAMCR.GT.0.) GO TO 52
   51 GAMCR=45.
   52 IF(DEL.GT.0.) GO TO 55
      DEL=.00001
   55 IF(ZETA.GT.0.) GO TO 53
      ZETA=.1E-30
   53 XKDB=1.
   54 CONTINUE
      READ(5,901)(B(I),I=1,K)
      READ(5,902)(FMT(I),I=1,12)
      READ(5,FMT)((X(I,L),L=1,M),I=1,N)
      READ(5,FMT)(Y(I),I=1,N)
      IWHER=-1
      GO TO 653
   59 IBKA=1
   58 WRITE(6,907),N,K,IP,M,IFP,GAMCR,DEL,FF,T,E,TAU,XL,ZETA
      IF(IFSS1.NE.1) GO TO 213
      WRITE(6,907)N,K,IP,M,IFP,GAMCR,DEL,FF,T,E,TAU,XL,ZETA
  213 CONTINUE
   60 CONTINUE
      DO 62 I=1,K
      G(I)=0.
      DO 62 J=1,K
   62 A(I,J)=0.
      GO TO (63,69,69),IBKA
   63 IF(IWS5.EQ.0) GO TO 630
      GO TO (66,64), IFSS3
  630 IFSS3=IWS3
      IFSS2=IWS2
      GO TO 70
   64 IFSS3=0
   66 GO TO (70,65),IFSS2
   65 IFSS2=0
      GO TO 70
   69 IFSS3=1
```

```
 70 WRITE(6,908)(B(J),J=1,K)
    IF(IFSS1.NE.1) GO TO 214
    WRITE(6,908)(B(J),J=1,K)
214 CONTINUE
    IF(IFSS3.RQ.0) GO TO 73
 68 WRITE (6,910)
    IF(IFSS1.NE.1) GO TO 259
    WRITE(6,910)
259 CONTINUE
 73 I=1
    PHI=0.
    IF(IFSS2.EQ.0) GO TO 57
    GO TO 600
 72 IF(IFSS2.EQ.1) GO TO 602
 57 IWHER=1
    GO TO 653
 75 IF(IP.LE.0) GO TO 80
 76 DO 77 II=1,IP
    IWS=IB(II)
 77 P(IWS)=0.
    GO TO 80
600 CONTINUE
602 IWHER=3
    GO TO 653
606 FWS=F
    DO 607 II=1,NPRNT
607 SPRNT(II)=PRNT(II)
    J=1
608 IF(IP.LE.0) GO TO 618
610 DO 612 II=1,IP
    IF((J-IB(II)).EQ.0) GO TO 621
612 CONTINUE
618 DBW=B(J)*DEL
    TWS=B(J)
    B(J)=B(J)+DBW
7002 IWHER=4
    GO TO 653
620 B(J)=TWS
    P(J)=(F-FWS)/DBW
    GO TO 622
621 P(J)=0.
622 J=J+1
    IF((J-K).LE.0) GO TO 608
624 F=FWS
    DO 625 II=1,NPRNT
625 PRNT(II)=SPRNT(II)
 80 DO 82 JJ=1,K
    G(JJ)=G(JJ)+(Y(I)-F)*P(JJ)
    DO 82 II=JJ,K
    VOS=P(II)*P(JJ)
    A(II,JJ)=A(II,JJ)+VOS
 82 A(JJ,II)=A(II,JJ)
318 WS=Y(I)-F
```

```
      IF(IFSS3.EQ.0)GO TO 314
  308 IF(NPRNT.GT.0) GO TO 312
  310 WRITE(6,925)Y(I),F,WS
      IF(IFSS1.NE.1) GO TO 219
      WRITE(6,925)Y(I),F,WS
  219 CONTINUE
      GO TO 314
  312 WRITE(6,925)Y(I),F,WS,(PRNT(JJ),JJ=1,NPRNT)
      IF(IFSS1.NE.1) GO TO 220
      WRITE(6,925)Y(I),F,WS,(PRNT(JJ),JJ=1,NPRNT)
  220 CONTINUE
  314 WS=Y(I)-F
      PHI=PHI+WS*WS
      I=I+1
      IF(I.LE.N) GO TO 72
   84 IF(IP.LE.0) GO TO 88
   85 DO 87 JJ=1,IP
      IWS=IB(JJ)
      DO 86 II=1,K
      A(IWS,II)=0.
   86 A(II,IWS)=0.
   87 A(IWS,IWS)=1.
   88 GO TO (90,704,706),IBKA
   90 DO 92 I=1,K
   92 SA(I)=SQRT(A(I,I))
      DO 106 I=1,K
      DO 100 J=1,K
      WS=SA(I)*SA(J)
      IF(WS.GT.0) GO TO 98
   96 A(I,J)=0.
      GO TO 100
   98 A(I,J)=A(I,J)/WS
  100 CONTINUE
      IF(SA(I).GT.0) GO TO 104
  102 G(I)=0.
      GO TO 106
  104 G(I)=G(I)/SA(I)
  106 CONTINUE
      DO 110 I=1,K
  110 A(I,I)=1.
      PHIZ=PHI
      GO TO (1132,1130),IBKT
 1130 CONTINUE
      GO TO 1134
 1132 DO 1133 II=1,K
      III=II+25
      DO 1133 JJ=1,K
 1133 A(III,JJ)=A(II,JJ)
 1134 CONTINUE
      IF(ITOT.GT.0) GO TO 163
  150 IF(XL.GT.0) GO TO 154
  152 XL=0.01
  154 ITOT=1
```

```
        DO 161 J=1,K
161 BS(J)=B(J)
163 IBK1=1
        WS=N-K+IP
        SE=SQRT(PHIZ/WS)
        IF(IFSS3.GT.0)GO TO 165
162 IF(IFSS2.EQ.0) GO TO 168
167 WRITE(6,911)PHIZ,SE,XLL,GAMMA,XL
221 CONTINUE
        GO TO 169
168 WRITE(6,912)PHIZ,SE,XLL,GAMMA,XL
        IF(IFSS1.NE.1)GO TO 222
        WRITE(6,912)PHIZ,SE,XLL,GAMMA,XL
222 CONTINUE
        IS=IS+1
        IF(IS.EQ.10) GO TO 881
891 GO TO 882
881 IS=0
        JS=JS+1
        IF(JS.EQ.1) GO TO 886
        IF(JS.EQ.10) GO TO 998
883 IF(PHIZ.GT.PHIA) GO TO 998
        PHIA= PHIZ
        GO TO 882
998 WRITE(6,985)
985 FORMAT(1H 7HDIVERGE)
        GO TO 6500
886 PHIA=PHIZ
882 IF(IFSS1.NE.1) GO TO 221
        GO TO 169
165 IF(IFSS2.EQ.0) GO TO 661
166 WRITE(6,903)PHIZ,SE,XL
        IF(IFSS1.NE.1) GO TO 223
        WRITE(6,903)PHIZ,SE,XL
223 CONTINUE
        GO TO 169
1661 WRITE (6,909)PHIZ,SE,XL
        IF(IFSS1.NE.1) GO TO 224
        WRITE(6,909)PHIZ,SE,XL
224 CONTINUE
169 GO TO 200
164 PHIL=PHI
        DO 170 J=1,K
        IF(ABS(DB(J)/(ABS(B(J))+TAU)).GE.E) GO TO 172
170 CONTINUE
        WRITE(6,923)
        IF(IFSS1.NE.1) GO TO 225
        WRITE(6,923)
225 CONTINUE
        GO TO 700
172 IF(IWS5.EQ.0) GO TO 1720
        GO TO (171,173),IFSS4
1720 IF(IWS4.EQ.0) GO TO 173
```

```
      IF(IWS4.EQ.1) GO TO 171
      IWS4=IWS4-1
      GO TO 173
  171 WRITE (6,924)
      IF(IFSS1.NE.1) GO TO 226
      WRITE(6,924)
  226 CONTINUE
      GO TO 700
  173 XKDB=1.
      IF(PHIL.GT.PHIZ) GO TO 190
  174 XLS=XL
      DO 176 J=1,K
      BA(J)=B(J)
  176 B(J)=BS(J)
      IF(XL.GT..00000001) GO TO 175
 1175 DO 1176J=1,K
      B(J)=BA(J)
 1176 BS(J)=B(J)
      GO TO 60
  175 XL=XL/10
      IBK1=2
      GO TO 200
  177 PHL4=PHI
      IF(PHL4.GT.PHIZ) GO TO 184
  182 DO 183 J=1,K
  183 BS(J)=B(J)
      GO TO 60
  184 XL=XLS
      DO 186 J=1,K
      BS(J)=BA(J)
  186 B(J)=BA(J)
      GO TO 60
  190 IBK1=4
      XLS=XL
      XL=XL/10.
      DO 185 J=1,K
  185 B(J)=BS(J)
      GO TO 200
  187 IF(PHI.LE.PHIZ) GO TO 196
  191 XL=XLS
      IBK1=3
  192 XL=XL*10.
  195 DO 193 J=1,K
  193 B(J)=BS(J)
      GO TO 200
  194 PHIT4=PHI
  180 IF(PHIT4.GT.PHIZ) GO TO 198
  196 DO 197 J=1,K
  197 BS(J)=B(J)
      GO TO 60
  198 IF(GAMMA.GE.GAMCR) GO TO 192
  199 XKDB=XKDB/2.
      DO 1199 J=1,K
```

```
      IF(ABS(DB(J)/(ABS(B(J))+TAU)).GE.E)GO TO 195
 1199 CONTINUE
      DO 1200 J=1,K
 1200 B(J)=BS(J)
      WRITE(6,934)
      IF(IFSS1.NE.1) GO TO 227
      WRITE (6,934)
  227 CONTINUE
      GO TO 700
  200 GO TO (1102,1100),IBKT
 1100 CONTINUE
      GO TO 1104
 1102 DO 1103 II=1,K
      III=II+25
      DO 1103 JJ=1,K
 1103 A(II,JJ)=A(III,JJ)
 1104 DO 202I=1,K
      A(I,I)=A(I,I)+XL
  202 CONTINUE
      IBKM=1
  404 CALL GJR(A,K,ZETA,MSING)
      GO TO (415,650),MSING
  415 GO TO (416,710),IBKM
  416 DO 420 I=1,K
      DB(I)=0.
      DO 421 J=1,K
  421 DB(I)=A(I,J)*G(J)+DB(I)
  420 DB(I)=XKDB*DB(I)
      XLL=0.
      DTG=0.
      GTG=0.
      DO 250 J=1,K
      DB(J)=DB(J)/SA(J)
      DTG=DTG+DB(J)*G(J)
      GTG=GTG+G(J)**2
      B(J)=B(J)+DB(J)
  250 XLL=XLL+DB(J)*DB(J)
      KIP=K-IP
      IF(KIP.EQ.1)GO TO 1257
      CGAM=DTG/SQRT(XLL*GTG)
      JGAM=1
      IF(CGAM.GT..0) GO TO 253
  251 CGAM=ABS(CGAM)
      JGAM=2
  253 GAMMA=57.2957795*(1.5707288+CGAM*(-0.212144+CGAM*(0.074261
     1-CGAM*.0187293)))*SQRT(1.-CGAM)
      GO TO (257,255),JGAM
  255 GAMMA=180.-GAMMA
      IF(XL.LT.0) GO TO 257
 1255 WRITE(6,922)XL,GAMMA
      IF(IFSS1.NE.1) GO TO 228
      WRITE(6,922)XL,GAMMA
  228 CONTINUE
      GO TO 700
```

```
1257 GAMMA=0.
 257 XLL=SQRT(XLL)
     IBK2=1
     GO TO 300
 252 IF(IFSS3.EQ.0) GO TO 256
 254 WRITE (6,904)(DB(J),J=1,K)
     IF(IFSS1.NE.1) GO TO 229
     WRITE(6,904)(DB(J),J=1,K)
 229 CONTINUE
     WRITE(6,905)PHI,XL,GAMMA,XLL
     IF(IFSS1.NE.1) GO TO 230
     WRITE(6,905)PHI,XL,GAMMA,XLL
 230 CONTINUE
 256 GO TO (164,177,194,187),IBK1
 300 I=1
     PHI=0.
     IWHER=2
     IF(IWS5.EQ.0)GO TO 653
     IF(NONSK.EQ.1) GO TO 650
 302 GO TO 653
 304 PHI=PHI+(Y(I)-F)**2
     I=I+1
     IF(I.LE.N)GO TO 302
 316 GO TO (252,780,704,762,766,772),IBK2
 700 DO 702 J=1,K
 702 B(J)=BS(J)
     WRITE(6,933)N,K,IP,M,FF,T,E,TAU
     IF(IFSS1.NE.1) GO TO 231
     WRITE(6,933)N,K,IP,M,FF,T,E,TAU
 231 CONTINUE
     IBKA=2
     GO TO 60
 704 IF(IFP.LE.0) GO TO 706
 705 IBKA=3
     IFP=0
     GO TO 60
 706 WS=N-K+IP
     SE=SQRT(PHI/WS)
     PHIZ=PHI
     IF(IFSS2.EQ.0)GO TO 709
 707 WRITE(6,903)PHIZ,SE,XL
     IF(IFSS1.NE.1) GO TO 232
     WRITE(6,903)PHIZ,SE,XL
 232 CONTINUE
     GO TO 708
 709 WRITE(_,909)PHIZ,SE,XL
     IF(IFSS1.NE.1) GO TO 233
     WRITE(6,909)PHIZ,SE,XL
 233 CONTINUE
 708 GO TO (1122,1120),IBKT
1120 CONTINUE
     GO TO 1124
1122 DO 1123 II=1,K
```

```
          III=II+25
          DO 1123 JJ=1,K
 1123 A(III,JJ)=A(II,JJ)
 1124 IBKM=2
          GO TO 404
  710 DO 711 J=1,K
          IF(A(J,J).LT..0) GO TO 713
  711 SA(J)=SQRT(A(J,J))
          GO TO 715
  713 IBOUT=1
  715 KST=-4
          WRITE(6,916)
          IF(IFSS1.NE.1) GO TO 234
          WRITE(6,916)
  234 KST=KST+5
          KEND=KST+4
          IF(KEND.LT.K) GO TO 719
          KEND=K
  719 DO 712 I=1,K
          IF(IFSS1.NE.1) GO TO 235
          WRITE(6,918)I,(A(I,J),J=KST,KEND)
  235 CONTINUE
  712 WRITE(6,918)I,(A(I,J),J=KST,KEND)
          IF(KEND.LT.K) GO TO 234
          IF(IBOUT.EQ.0) GO TO 717
          WRITE(6,936)
          IF(IFSS1.NE.1) GO TO 6500
          WRITE(6,936)
          GO TO 6500
  717 DO 718 I=1,K
          DO 718 J=1,K
          WS=SA(I)*SA(J)
          IF(WS.GT.0.) GO TO 716
  714 A(I,J)=0.
          GO TO 718
  716 A(I,J)=A(I,J)/WS
  718 CONTINUE
          DO 720 J=1,K
  720 A(J,J)=1.
          WRITE(6,917)
          IF(IFSS1.NE.1) GO TO 236
          WRITE(6,917)
  236 CONTINUE
          KST=-9
  721 KST=KST+10
          KEND=KST+9
          IF(KEND.LT.K) GO TO 722
          KEND=K
  722 DO 724 I=1,K
          IF(IFSS1.NE.1) GO TO 237
          WRITE(6,935)I,(A(I,J),J=KST,KEND)
  237 CONTINUE
  724 WRITE(6,935)I,(A(I,J),J=KST,KEND)
```

```
        IF(KEND.LT.K) GO TO 721
        DO 726 J=1,K
  726 SA(J)=SE*SA(J)
        GO TO (1112,1110),IBKT
 1110 CONTINUE
        GO TO 1114
 1112 DO 1113 II=1,K
        III=II+25
        DO 1113 JJ=1,K
 1113 A(II,JJ)=A(III,JJ)
 1114 CONTINUE
  740 WRITE (6,919)
        IF(IFSS1.NE.1) GO TO 238
        WRITE(6,919)
  238 CONTINUE
        WS=K-IP
        DO 750 J=1,K
        IF(IP.LE.0) GO TO 743
  741 DO 742 I=1,IP
        IF(J.EQ.IB(I)) GO TO 746
  742 CONTINUE
  743 HJTD=SQRT(WS*FF)*SA(J)
        STE=SA(J)
        OPL=BS(J)-SA(J)*T
        OPU=BS(J)+SA(J)*T
        SPL=BS(J)-HJTD
        SPU=BS(J)+HJTD
        WRITE(6,927)J,STE,OPL,OPU,SPL,SPU
        IF(IFSS1.NE.1) GO TO 239
        WRITE(6,927)J,STE,OPL,OPU,SPL,SPU
  239 CONTINUE
        GO TO 750
  746 WRITE(6,913)J
        IF(IFSS1.NE.1) GO TO 240
        WRITE(6,913)J
  240 CONTINUE
  750 CONTINUE
        IF(IWS6.EQ.1) GO TO 650
        WS=K-IP
        WS1= N-K+IP
        PKN=WS/WS1
        PC=PHIZ*(1.+FF*PKN)
        WRITE(6,920)PC
        IF(IFSS1.NE.1) GO TO 241
        WRITE(6,920)PC
  241 CONTINUE
        WRITE(6,921)
        IF(IFSS1.NE.1) GO TO 242
        WRITE(6,921)
  242 CONTINUE
        IFSS3=1
        J=1
 7790 IBKP=1
```

```
      IBKP=1
      DO 752 JJ=1,K
752   B(JJ)=BS(JJ)
      IF(IP.LE.0) GO TO 758
754   DO 756 JJ=1,IP
      IF(J.EQ.IB(JJ)) GO TO 787
756   CONTINUE
758   DD=-1
      IBKN=1
760   D=DD
      B(J)=BS(J)+D*SA(J)
      IBK2=4
      GO TO 300
762   PHI1=PHI
      IF(PHI1.GE.PC) GO TO 770
764   D=D+DD
      IF(D/DD.GE.5) GO TO 788
765   B(J)=BS(J)+D*SA(J)
      IBK2=5
      GO TO 300
766   PHID=PHI
      IF(PHID.LT.PC) GO TO 764
      IF(PHID.GE.PC) GO TO 778
770   D=D/2.
      IF(D/DD.LE..001) GO TO 788
771   B(J)=BS(J)+D*SA(J)
      IBK2=6
      GO TO 300
772   PHID=PHI
      IF(PHID.GT.PC) GO TO 770
778   XK1=PHIZ/D+PHI1/(1/-D)+PHID/(D*(D-1.))
      XK2=-(PHIZ*(1.+D)/D+D/(1.-D)*PHI1+PHID/(D*(D-1.)))
      XK3=PHIZ-PC
      BC=(SQRT(XK2*XK2-4.*XK1*XK3)-XK2)/(2.*XK1)
      GO TO (779,784),IBKN
779   B(J)=BS(J)-SA(J)*BC
      GO TO 781
784   B(J)=BS(J)+SA(J)*BC
781   IBK2=2
      GO TO 300
780   GO TO (782,786),IBKN
782   IBKN=2
      DD=1.
      BL=B(J)
      PL=PHI
      GO TO 760
786   BU=B(J)
      PU=PHI
      GO TO (783,795,785,789),IBKP
783   WRITE(6,918)J,BL,PL,BU,PU
      IF(IFSS1.NE.1) GO TO 243
      WRITE(6,918)J,BL,PL,BU,PU
243   CONTINUE
```

```
      GO TO 790
795   WRITE(6,915)J,BU,PU
      IF(IFSS1.NE.1) GO TO 244
      WRITE(6,915)J,BU,PU
244   CONTINUE
      GO TO 790
785   WRITE(6,918)J,BL,PL
      IF(IFSS1.NE.1) GO TO 245
      WRITE(6,918)J,BL,PL
245   CONTINUE
      GOTO 790
787   WRITE(6,913)J
      IF(IFSS1.NE.1) GO TO 246
      WRITE(6,913)J
246   CONTINUE
      GO TO 790
789   WRITE(6,914)J
      IF(IFSS1.NE.1) GO TO 247
      WRITE(6,914)J
247   CONTINUE
      GO TO 790
788   GO TO (791,792),IBKN
791   IBKP=2
      GO TO 780
792   GO TO (793,794),IBKP
793   IBKP=3
      GO TO 780
794   IBKP=4
      GO TO 780
790   J=J+1
      IF(J.LE.K) GO TO 790
      GO TO 6500
900   FORMAT(25I3)
901   FORMAT(7F10.0)
902   FORMAT(18A4)
903   FORMAT(1H 13X,4H PHI 14X,4H S E         9X,7H LAMBDA
     16X, 25H ESTIMATED PARTIALS USED  / 5X,2E18.8,  E13.3)
904   FORMAT(1H 12H increments 5E18.8/(12X,5E18.8))
905   FORMAT(1H 13X,4H PHI 10X,7H LAMBDA 6X,7H GAMMA  6X,7HLENGTH /
     1 5X, E18.8, 3E13.3)
906   FORMAT(1H ,1E9.2,86X,1E9.2 /1X,1HN 99X,1HN )
907   FORMAT( 5H1N = I3,5X,5H K = I3,5X,5H P = I3,5X,5H m = I3,5X,
     1 7H IFP = I3,5X,13HGAMMA CRIT = E10.3,5X,5HDEL = E10.3/6H ff =
     2E10.3,5X,5H T = E10.3,5X,5H E = E10.3,5X,7H TAU = E10.3,5X,6H
     3 XL = E10.3, 4X, 7HZETA = E10.3 /)
908   FORMAT(1H 12H PARAMETERS 5E18.8/(12X,5E18.8) )
909   FORMAT(1H 13X,4H PHI 14X, 4H S E         9X,7H LAMBDA 6X,
     1 25H ANALYTIC PARTIALS USED    /5X, 2E18.8  E13.3)
910   FORMAT(1H /5X,9X,4H OBS 13X,5H PRED 13X,5H DIFF  )
911   FORMAT(1H 13X,4H PHI 14X,4H S E 11X,7H LENGTH 6X, 7H GAMMA 6X,
     1 7H LAMBDA 6X, 25 HESTIMATED PARTIALS USED /5X, 2E18.8, 3E13.3)
912   FORMAT(1H 13X,4H PHI 14X,4H S E 11X,7H LENGTH 6X,  7H GAMMA 6X
     1,7HLAMBDA 6X, 24HANALYTIC PARTIALS USED  /5X,2E18.8,3E13.3)
```

```
 913 FORMAT( 1H 2X,I3,20H PARAMETER NOT USED   )
 914 FORMAT( 1H 2X,I3,12H NONE FOUND   )
 915 FORMAT( 1H 2X,I3,36X,2E18.8)
 916 FORMAT( 1H /13H PTP INVERSE )
 917 FORMAT( 1H /30H PARAMETER CORRELATION MATRIX   )
 918 FORMAT( 2X,I3,5E18.8)
 919 FORMAT( 1H /1H / 13X,4H STD 17X, 16H ONE - PARAMETER 21X,
     1 14H SUPPORT PLANE / 3X, 2H B 7X,6H ERROR 12X, 6H LOWER
     2 12X, 6H UPPER 12X, 6H LOWER 12X, 6H UPPER   )
 920 FORMAT( 1H /1H /30H NONLINEAR CONFIDENCE LIMITS  / /
     1 16H PHI CRITICAL = E15.8   )
 921 FORMAT( 1H / 6H PARA 6X,8H LOWER B 8X,10H LOWER PHI 10X,8H
     1 UPPER B 8X,10H UPPER PHI   )
 922 FORMAT( 18H GAMMA LAMBDA TEST,5X,2E13.3)
 923 FORMAT( 14H EPSILON TEST   )
 924 FORMAT( 11H FOURCE OFF   )
 925 FORMAT( 1H 5X,6E18.8/59X,2E18.8/)
 926 FORMAT( 40H BAD DATA, SUBSCRIPTS FOR UNUSED BS = 0   / / / )
 927 FORMAT( 1H 2X,I3,5E18.8)
 928 FORMAT( 1H ,110A1   )
 929 FORMAT( 1H 10A1)
 931 FORMAT( 8F10.0)
 932 FORMAT( 1H1)
 933 FORMAT( 5H1N = I3,5X,5H K = I3,5X,5H P = I3,5X,5H M = I3,5X,
     1                         / 6H FF = E10.3,5X,5H T = E
     2 10.3,5X,5H E = E10.3,5X,7H TAU = E10.3   /   )
 934 FORMAT( 19H GAMMA EPSILON TEST   )
 935 FORMAT( 1H 3X,I5,2X,10F10.4)
 936 FORMAT( 27H0 NEGATIVE DIAGONAL ELEMENT   )
 999 FORMAT( 2(E20.10))
1001 FORMAT( E20.10)
1002 FORMAT( 1H ,3E12.5)
1003 FORMAT( 1H ,7E12.5,I3)
1009 FORMAT( 1H ,7E12.5)
     END

     SUBROUTINE GJR(A,N,EPS,MSING)
     DIMENSION A(20,21),B(20),C(20),P(20),Q(20)
     INTEGER P,Q
     MSING=1
     DO 10 K=1,N
     PIVOT=0
     DO 20 I=K,N
     DO 20 J=K,N
     IF(ABS(A(I,J))-ABS(PIVOT))20,20,30
  30 PIVOT=A(I,J)
     P(K)=I
     Q(K)=J
  20 CONTINUE
     IF(ABS(PIVOT)-EPS)40,40,50
  50 IF(P(K)-K)60,80,60
  60 DO 70 J=1,N
```

```
          L=P(K)
          Z=A(L,J)
          A(L,J)=A(K,J)
   70 A(K,J)=Z
   80 IF(Q(K)-K)85,90,85
   85 DO 100 I=1,N
          L=Q(K)
          Z=A(I,L)
          A(I,L)=A(I,K)
  100 A(I,K)=Z
   90 CONTINUE
          DO 110 J=1,N
          IF(J-K)130,120,130
  120 B(J)=1./PIVOT
          C(J)=1.
          GO TO 140
  130 B(J)=-A(K,J)/PIVOT
          C(J)=A(J,K)
  140 A(K,J)=0.
  110 A(J,K)=0.
          DO 10 I=1,N
          DO 10 J=1,N
   10 A(I,J)=A(I,J)+C(I)*B(J)
          DO 155 M=1,N
          K=N-M+1
          IF(P(K)-K)160,170,160
  160 DO 180 I=1,N
          L=P(K)
          Z=A(I,L)
          A(I,L)=A(I,K)
  180 A(I,K)=Z
  170 IF(Q(K)-K)190,155,190
  190 DO 150 J=1,N
          L=Q(K)
          Z=A(L,J)
          A(L,J)=A(K,J)
  150 A(K,J)=Z
  155 CONTINUE
  151 RETURN
   40 PRINT45,P(K),Q(K),PIVOT
   45 FORMAT(16HOSINGULAR MATRIX3H I=I3,3H J=I3,7H PIVOT=E16.8/)
          MSING=2
          GO TO 151
          END

          SUBROUTINE SUBZ(Y,X,B,PRNT,NPRNT,N)
          DIMENSION Y(100),X(100,10),B(20),PRNT(5)
          NPRNT=1.
          RETURN
          END
```

```
SUBROUTINE FCODE(Y,X,B,PRNT,F,I)
DIMENSION Y(100),X(100,10),B(20),PRNT(5)
F=B(2)-B(1)*ALOG(1.-(X(I,1)))
PRNT(1)=X(I,1)
RETURN
END

SUBROUTINE PCODE(P,X,B,PRNT,F,I)
DIMENSION P(20),X(100,10),B(20),PRNT(5)
P(1)=-ALOG(1.-(X(I,1)))
P(2)=1.
RETURN
END
```

```
/&
// EXEC LNKEDT
// EXEL
```

REFERENCES[a]

Acker, L. (1969). Water activity and enzyme activity. *Food Technol.* **23**: 1257. [1]

Agrawal, K. K., Clary, B. L., and Nelson, G. L. (1969). Investigation into the theories of desorption isotherms for rough rice and peanuts. I. Paper No. 69-890, 1969 Winter Meeting American Society of Agricultural Engineers, Chicago, Illinois. [2]

Alcaraz, E. C., Martín, M. A., and Marín, J. P. (1977). Método manométrico para medida de humedades de equilibrio. *Grasas y Aceites* **28**: 403. [3]

Audu, T. O. K., Loncin, M., and Weisser, H. (1978). Sorption isotherms of sugars. *Lebensm. Wiss. Technol.* **11**: 31. [4]

Ayernor, G. S., and Steinberg, M. P. (1977). Hydration and rheology of soy-fortified pregelled corn flours. *J. Food Sci.* **42**: 65. [5]

Ayerst, G. (1965). Determination of the water activity of some hygroscopic food materials by a dew-point method. *J. Sci. Food Agric.* **16**: 71. [6]

Baisya, R. K., Chattoraj, D. K., and Bose, A. N. (1975). Water binding characteristics of normal curd (Dahi) and rehydrated curd powder gels. *Indian J. Technol.* **13**: 578. [7]

Bakharev, I. Ya. (1960). Quoted by Agrawal *et al.* (1969). [8]

Baloch, A. K., Buckle, K. A., and Edwards, R. A. (1977). Stability of β-carotene in model systems containing sulphite. *J. Food Technol.* **12**: 309. [9]

Bandyopadhyay, S., Weisser, H., and Loncin, M. (1980). Water adsorption isotherms of foods at high temperatures. *Lebensm. Wiss. Technol.* **13**: 182. [9a]

Barron, L. F. (1977). The expansion of wafer and its relation to the cracking of chocolate and "bakers" chocolate coatings. *J. Food Technol.* **12**: 73. [10]

Barthölomai, G. B., Brennan, J. G., and Jowitt, R. (1975). Mechanisms of volatile retention in freeze-dried food liquids. *Lebensm. Wiss. Technol.* **8**: 25. [11]

Beasly, E. O. (1962). Moisture equilibrium of Virginia Bunch peanuts. MS. thesis, North Carolina State University, Raleigh. [12]

Becker, H. A., and Sallans, H. R. (1956). A study of the desorption isotherms of wheat at 25°C and 50°C. *Cereal Chem.* **33**: 79. [13]

Berlin, E., and Anderson, B. A. (1975). Reversibility of water vapor sorption by cottage cheese whey solids. *J. Dairy Sci.* **58**: 25. [19]

Berlin, E., Anderson, B. A., and Pallansch, M. J. (1968a). Water vapor sorption properties of various dried milks and wheys. *J. Dairy Sci.* **51**: 1339. [14]

Berlin, E., Anderson, B. A., and Pallansch, M. J. (1968b). Comparison of water vapor sorption by milk powder components. *J. Dairy Sci.* **51**: 1912. [15]

Berlin, E., Anderson, B. A., and Pallansch, M. J. (1970). Effect of temperature on water vapor sorption by dried milk powders. *J. Dairy Sci.* **53**: 146. [16]

Berlin, E., Kliman, P. G., Anderson, B. A., and Pallansch, M. J. (1973a). Water binding in whey protein concentrates. *J. Dairy Sci.* **56**: 984. [17]

Berlin, E., Anderson, B. A., and Pallansch, M. J. (1973b). Water sorption by dried dairy products stabilized with carboxymethyl cellulose. *J. Dairy Sci.* **56**: 685. [18]

Bolin, H. R. (1980). Relation of moisture to water activity in prunes and raisins. *J. Food Sci.* **45**: 1190. [18a]

Boquet, R., Chirife, J., and Iglesias, H. A. (1978). Equations for fitting water sorption isotherms of foods. Part II—Evaluation of various two-parameter models. *J. Food Technol.* **13**: 319. [20]

[a] Bold numbers within brackets refer to the references cited in Tables I and II.

Bradley, R. S. (1936). *J. Chem. Soc.*: 1467. [21]

Breese, M. H. (1955). Hysteresis in the hygroscopic equilibria of rough rice at 25°C. *Cereal Chem.* 32: 481. [22]

Brunauer, S., Emmet, P. H., and Teller, E. (1938). Adsorption of gases in multimolecular layers. *J. Am. Chem. Soc.* 60: 309. [23]

Bull, H. B. (1944). Adsorption of water vapor by proteins. *J. Am. Chem. Soc.* 66: 1499. [24]

Bushuk, W., and Winkler, C. A. (1957). Sorption of water vapor on wheat flour, starch and gluten. *Cereal Chem.* 34: 73. [25]

Chen, C. S. (1971). Equilibrium moisture curves for biological materials. *Trans. ASAE* 14: 924. [26]

Chen, C. S., and Clayton, J. T. (1971). The effect of temperature on sorption isotherms of biological materials. *Trans. ASAE* 14: 927. [27]

Chilton, W. G., and Collison, R. (1974). Hydration and gelation of modified potato starches. *J. Food Technol.* 9: 87. [28]

Chirife, J., and Iglesias, H. A. (1978). Equations for fitting water sorption isotherms of foods: Part 1 — A review. *J. Food Technol.* 13: 159. [30]

Chirife, J., and Karel, M. (1974). Effect of structure disrupting treatments on volatile release from freeze-dried maltose. *J. Food Technol.* 9: 13. [29]

Christian, J. H. B., and Waltho, J. A. (1962). The water relations of staphylococci and micrococci. *J. Appl. Bacteriol.* 25: 369. [31]

Chung, D. S., and Pfost, H. B. (1967). Adsorption and desorption of water vapor by cereal grains and their products. *Trans. ASAE* 10: 549. [32]

Coleman, D. A., and Fellows, H. C. (1925). Hygroscopic moisture of cereal grains and flaxseed exposed to atmospheres of different relative humidities. *Cereal Chem.* 2: 275. [33]

Cooper, R. M., Knight, R. A., Robb, J., and Seiler, D. A. L. (19). Equilibrium relative humidity of cakes. *Food Trade Rev.* 38: 40. [34]

Delaney, R. A. M. (1977). Protein concentrates from slaughter animal blood. II. Composition and properties of spray dried red blood cell concentrates. *J. Food Technol.* 12: 355. [35]

Denizel, T., Rolfe, E. J., and Jarvis, B. (1976). Moisture–equilibrium relative humidity relationships in pistacho nuts with particular regard to control of aflatoxin formation. *J. Sci. Food Agric.* 27: 1027. [36]

Dexter, S. T., Andersen, A. L. Pfahler, P. L., and Benne, E. J. (1955). Responses of white pea beans to various humidities and temperature of storage. *Agron. J.* 57: 246. [37]

Ditmar, J. H. (1935). Hygroscopicity of sugars and sugar mixtures. *Ind. Eng. Chem.* 27: 333. [38]

Duckworth, R. B. (1972). The properties of water around the surfaces of food colloids. *Proc. Inst. Food Sci. Technol.* (U.K.) 5: 60. [39]

Duckworth, R. B., and Smith, G. M. (1963). The environment for chemical change in dried and frozen foods. *Proc. Nutr. Soc.* 22: 182. [40]

Eichner, K. (1975). The influence of water content on nonenzymic browning reactions in dehydrated foods and model systems and the inhibition of fat oxidation by browning intermediates. *In* "Water Relations of Foods" (R. B. Duckworth, ed.), p. 417. Academic Press, New York. [41]

Eichner, K., and Ciner-Doruk, M. (1979). Bildung und stabilität von Amadori-verbindungen in wasserarmen modellsystemen. *Z. Lebensm. Unters. Forsch.* 168: 360. [42]

Fenton, F. C. (1941). Storage of grain sorghums. *Agric. Eng.* 22: 185. [43]

Ferrel, R. E., Shepherd, A. D., Thielking, R., and Pence, J. W. (1966). Moisture equilibrium of bulgur. *Cereal Chem.* 43: 136. [44]

Fish, B. P. (1958). Diffusion and thermodynamics of water in potato starch. *In* "Fundamental Aspects of the Dehydration of Foodstuff," p. 143. Macmillan, New York. [45]

Flink, J. M., and Karel, M. (1972). Mechanisms of retention of organic volatiles in freeze-dried systems. *J. Food Technol.* 7: 199. [46]

Gál, S. (1975). Recent advances in techniques for the determination of sorption isotherms. *In* "Water Relations of Foods" (R. B. Duckworth, ed.), p. 139. Academic Press, New York. **[47]**

Gane, R. (1943). Dried egg. VI. The water relations of dried egg. *J. Soc. Chem. Ind.* **42:** 185. **[48]**

Gane, R. (1948). The water content of the seeds of peas, soybeans, linseed, grass, onion and carrot as a function of temperature and humidity of the atmosphere. *J. Agric. Sci.* **38**(1): 81. **[49]**

Gane, R. (1950). The water relations of some dried fruits, vegetables and plant products. *J. Sci. Food Agric.* **1:** 42. **[50]**

Geurts, T. J., Walstra, P., and Mulder, H. (1974). Water binding to milk protein, with particular reference to cheese. *Neth. Milk Dairy J.* **28:** 46. **[51]**

Greig, R. I. W. (1979). Sorption properties of heat denatured cheese whey protein. Part II. Unfreezable water content. *Dairy Industries International,* June 1979, p. 15. **[51a]**

Gustafson, R. J., and Hall, G. E. (1974). Equilibrium moisture content of shelled corn from 50 to 155°F. *Trans. ASAE* **17:** 120. **[52]**

Gur-Arieh, C., Nelson, A. I., Steinberg, M. P., and Wei, L. S. (1967). Moisture adsorption by wheat flours and their cake baking performance. *Food Techn.* **21:** 412. **[52a]**

Halsey, G. (1948). Physical adsorption in non-uniform surfaces *J. Chem. Phys.* **16:** 931. **[53]**

Hansen, J. R. (1976). Hydration of soybean protein. *J. Agric. Food Chem.* **24:** 1136. **[54]**

Harel, S., Kranner, J., Juven, B. J., and Golan, R. (1978). Long-term preservation of high-moisture dried apricots with and without chemical preservatives. *Lebensm. Wiss. Technol.* **11:** 219. **[55]**

Hayakawa, K. I., Matas, J., and Hwang, P. (1978). Moisture sorption isotherms of coffee products. *J. Food Sci.* **43:** 1026. **[56]**

Heldman, D. R., Hall, C. W., and Hedrick, T. I. (1965). Vapor equilibrium relationships of dry milk. *J. Dairy Sci.* **48:** 845. **[57]**

Henderson, S. M. (1952). A basic concept of equilibrium moisture. *Agric. Eng.* **33:** 29. **[58]**

Henderson, S. M. (1969). Equilibrium moisture content of small grain hysteresis. ASAE paper No. 69-329. ASAE, St. Joseph, Michigan. **[59]**

Henderson, S. M. (1973). Equilibrium moisture content of hops. *J. Agric. Eng. Res.* **18:** 55. **[60]**

Hermansson, A. M. (1977). Functional properties of proteins for foods—water vapour sorption. *J. Food Technol.* **12:** 177. **[61]**

Hogan, J. T., and Karon, M. L. (1955). Hygroscopic equilibria of rough rice at elevated temperatures. *Agric. Food Chem.* **3:** 855. **[62]**

Hubbard, J. E., Earle, F. R., and Senti, F. R. (1957). Moisture relations in wheat and corn. *Cereal Chem.* **34:** 422. **[63]**

Iglesias, H. A. (1975). Isotermas de sorción de agua en remolacha azucarera y análisis del fenómeno de sorción en alimentos. Thesis, Departamento de Industrias, Facultad de Ciencias Exactas y Naturales, Universidad de Buenos Aires, Argentina. **[64]**

Iglesias, H. A., and Chirife, J. (1976a). Isosteric heats of water vapour sorption in dehydrated foods. Part 1: Analysis of the differential heat curves. *Lebensm. Wiss. Technol.* **9:** 107. **[68]**

Iglesias, H. A., and Chirife, J. (1976b). Equilibrium moisture content of air-dried beef. Dependence on drying temperature. *J. Food Technol.* **11:** 565. **[69]**

Iglesias, H. A., and Chirife, J. (1976c). B.E.T. monolayer values in dehydrated foods and food components. *Lebensm. Wiss. Technol.* **9:** 107. **[70]**

Iglesias, H. A., and Chirife, J. (1976d). Prediction of the effect of temperature on water sorption isotherms of food material. *J. Food Technol.* **11:** 109. **[70a]**

Iglesias, H. A., and Chirife, J. (1977a). Effect of fat content on the water sorption isotherm of air dried minced beef. *Lebensm. Wiss. Technol.* **10:** 151. **[70b]**

Iglesias, H. A., and Chirife, J. (1977b). Effect of heating in the dried state on the water sorption by beef. *Lebensm. Wiss. Technol.* **10:** 249. **[72]**

Iglesias, H. A., and Chirife, J. (1978). An empirical equation for fitting water sorption isotherms of fruits and related products. *Can. Inst. Food Sci. Technol. J.* **11:** 12. **[74]**

Iglesias, H. A., and Chirife, J. (1981). An equation for fitting uncommon water sorption isotherms in foods. *Lebensm. Wiss. Technol.* **14**: 105. [76]

Iglesias, H. A., Chirife, J., and Lombardi, J. L. (1975a). Water sorption isotherms in sugar beet root. *J. Food Technol.* **10**: 299. [65]

Iglesias, H. A., Chirife, J., and Lombardi, J. L. (1975b). Comparison of water vapor sorption by sugar beet root components. *J. Food Technol.* **10**: 385. [66]

Iglesias, H. A., Chirife, J., and Lombardi, J. L. (1975c). An equation for correlating equilibrium moisture content in foods. *J. Food Technol.* **10**: 289. [67]

Iglesias, H. A., Chirife, J., and Viollaz, P. E. (1976). Thermodynamics of water vapour sorption in sugar beet root. *J. Food Technol.* **11**: 91. [71]

Iglesias, H. A., Chirife, J., and Viollaz, P. E. (1977). Evaluation of some factors useful for the mathematical prediction of moisture gain by packaged dried beef. *J. Food Technol.* **12**: 505. [72]

Iglesias, H. A., Viollaz, P. E., and Chirife, J. (1979). A technique for predicting moisture transfer in mixtures of dehydrated foods. *J. Food Technol.* **14**: 89. [73]

Iglesias, H. A., Chirife, J., and Boquet, R. (1980). Prediction of water sorption isotherms of food models from knowledge of components sorption behaviour. *J. Food Sci.* **45**: 450. [77]

Jason, A. C. (1958). A study of evaporation and diffusion processes in the drying of fish muscle. In "Fundamental Aspects of the Dehydration of Foodstuffs," p. 103. MacMillan, New York. [78]

Jayaratnam, S., and Kirtisinghe, D. (1974a). The effect of relative humidity and temperature on moisture sorption by black tea. *Tea Quart.* **44**: 164. [79]

Jayaratnam, S., and Kirtisinghe, D. (1974b). The effect of relative humidity on the storage life of made tea. *Tea Quart.* **44**: 170. [80]

Jordão, B. A., and Stoff, S. R. (1969–70). Curva de saturacão do feijão de mesa variedade rosinha. *Coletanea Inst. Tecnol. Alimentos,* Campinas, Brazil. **3**: 425. [81]

Juliano, B. O. (1964). Hygroscopic equilibria of rough rice. *Cereal Chem.* **41**: 191. [82]

Kapsalis, J. G. (1975). The influence of water on textural parameters in foods at intermediate moisture levels. In "Water Relations of Foods" (R. B. Duckworth, ed.), p. 627. Academic Press, New York. [84]

Kapsalis, J. G., Wolf, M., Driver, M., and Walker, J. E. (1971). The effect of moisture on the flavor content and texture stability of dehydrated foods. *ASHRAE J.,* January, p. 93. [83]

Karel, M. (1973). Recent research in the area of low moisture and intermediate moisture foods. *Crit. Rev. Food Technol.:* 329. [87]

Karel, M., and Labuza, T. P. (1968). Nonenzymatic browning in model systems containing sucrose. *J. Agric. Food Chem.* **16**: 717. [86]

Karel, M., and Nickerson, J. T. R. (1964). Effects of relative humidity, air and vacuum on browning of dehydrated orange juice. *Food Technol.* **18**: 104. [85]

Kargin, X. (1957). Quoted by Smith (1947). [88]

Karon, M. L., and Adams, M. E. (1949). Hygroscopic equilibria of rice and rice fractions. *Cereal Chem.* **26**: 1. [90]

Karon, M. L., and Hillery, B. E. (1949). Hygroscopic equilibrium of peanuts. *J. Am. Oil Chem. Soc.* **26**: 16. [89]

King, C. J. (1968). Rates of moisture sorption and desorption in porous, dried foodstuffs. *Food Technol.* **22**: 509. [91]

King, C. J., Lam, W. K., and Sandall, O. C. (1968). Physical properties important for freeze drying poultry meat. *Food Technol.* **22**: 1302. [92]

Koizumi, C., Iiyama, S., Wada, S., and Nonaka, J. (1978). Lipid deteriorations of freeze-dried fish meats at different equilibrium relative humidities. *Bull. Japan. Soc. Sci. Fisheries* **44**: 209. [93]

Komeyasu, M., and Iyama, M. (1974). Studies on spray-dried citrus unshiu juice. Part 1. Characteristics of water adsorption properties of spray-dried citrus unshiu juice. *Nippon Shokuhin Kogyo Gakkaishi* **21**: 384. [94]

Kopelman, I. J., and Saguy, I. (1977). Drum dried beet powder. *J. Food Technol.* **12**: 615. [95]

Kopelman, I. J., Meydav, S., and Weinberg, S. (1977). Storage studies of freeze dried lemon crystals. *J. Food Technol.* **12**: 403. **[96]**

Koury, B. J., and Spinelli, J. (1975). Effect of moisture, carbohydrate and atmosphere on the functional stability of fish protein isolates. *J. Food Sci.* **40**: 58. **[97]**

Kumar, M. (1974). Water vapour adsorption on whole corn flour, degermed corn flour, and germ flour. *J. Food Technol.* **9**: 433. **[98]**

Kuprianoff, J. (1962). Some factors influencing the reversibility of freeze-drying of foodstuffs. *In* "Freeze Drying of Foods" (F. R. Fisher, ed.). National Academy of Sciences–National Research Council, Washington, D.C. **[99]**

Labuza, T. P. (1968). Sorption phenomena in foods. *Food Technol.* **22**: 263. **[100]**

Labuza, T. P. (1971). Kinetics of lipid oxidation in foods. *CRC Crit. Rev. Food Technol.*, October, p. 355. **[101]**

Labuza, T. P. (1975). Oxidative changes in foods at low and intermediate moisture levels. *In* "Water Relations of Foods" (R. B. Duckworth, ed.), p. 455. Academic Press, New York. **[104]**

Labuza, T. P., and Rutman, M. (1968). The effect of surface active agents on sorption isotherms of a model food system. *Can. J. Chem. Eng.* **46**: 364. **[102]**

Labuza, T. P., Tannenbaum, S. R., and Karel, M. (1970). Water content and stability of low moisture and intermediate-moisture foods. *Food Technol.* **24**: 543. **[105]**

Labuza, T. P., Mizrahi, S., and Karel, M. (1972). Mathematical models for optimization of flexible film packaging of foods for storage. *Trans. ASAE* **15**: 150. **[103]**

Labuza, T. P., Acott, K., Tatini, S. R., Lee, R. Y., and Flink, J. M. (1976). Water activity determination: A collaborative study of different methods. *J. Food Sci.* **41**: 911. **[106]**

Lafuente, B., and Piñaga, F. (1966). Humedades de equilibrio de productos liofilizados. *Rev. Agroq. Tecnol. Alimentos* **6**: 113. **[107]**

Landrock, A. H., and Proctor, B. E. (1951). A new graphical interpolation method for obtaining humidity equilibria data, with special reference to its role in food packaging studies. *Food Technol.* **5**: 332. **[108]**

Leistner, L., and Rödel, W. (1975). The significance of water activity for micro-organisms in meats. *In* "Water Relations of Foods" (R. B. Duckworth, ed.), p. 309. Academic Press, New York. **[109]**

Lewicki, P. P. (1976). Powtarzlnose statycznoeksykatorowej metody wyznaczania izoterm sorpcji wody produktów spozywczych. *Przemysl Spozywczy* **30**: 141. **[110]**

Lewicki, P. P., and Brzozowski, J. (1973). Badanie izoterm adsorpcji wody wybranych produktów spozywczych. *Przemysl Spozywczy* **27**: 18. **[111]**

Lewicki, P. P., and Lenart, A. (1975). Wptyw procesu technologicznego na wlaściwości adsorpcyjne marchwi i porów suszonych. *Przemysl Spozywczy* **29**: 73. **[112]**

Lewicki, P. P., and Lenart, A. (1977). Wptyw wstepnego odwadniania osmotycznego na wlásciwości adsorpcyjne jablek suszonych owiewowo. *Przemysl Spozywczy* **31**: 394. **[113]**

Lladser, M., and Piñaga, F. (1975). Criodeshidratación de aguacates. I. Estudio sobre el comportamiento eutéctico e higroscópico del aguacate liofilizado y ensayo de almacenamiento acelerado del mismo. *Rev. Agroq. Tecnol. Alimentos* **15**: 547. **[114]**

Loncin, M., Bimbenet, J. J., and Lenges, L. (1968). Influence of the activity of water on the spoilage of foodstuffs. *J. Food Technol.* **3**: 131. **[115]**

McBean, D. McG., and Wallace, J. J. (1967). Stability of moist-pack apricots in storage. *CSIRO Food Preservation Quart.* **27**: 29. **[122]**

McCurdy, A. R., Leung, H. K., and Swanson, B. G. (1980). Moisture equilibration and measurement in dry beans (*Phaseolus vulgaris*). *J. Food Sci.* **45**: 506. **[123]**

MacKenzie, A. P., and Luyet, B. J. (1967). Water sorption isotherms from freeze-dried muscle fibers. *Cryobiology* **3**: 341. **[116]**

Makower, B. (1945). Vapor pressure of water adsorbed on dehydrated eggs. *Ind. Eng. Chem.* **37**: 1018. **[117]**

Makower, B., and Dehority, G. L. (1943). Equilibrium moisture content of dehydrated vegetables. *Ind. Eng. Chem.* **35**: 193. **[118]**

Makower, B., and Dye, W. B. (1956). Equilibrium moisture content and crystallization of amorphous sucrose and glucose. *J. Agric. Food Chem.* **4**: 72. **[119]**

Malthlouthi, J. F., Michel, J. F., and Maitenaz, P. C. (1981). Study of some factors affecting water vapor sorption of gruyere cheese. I. Proteolysis. *Lebensm. Wiss. Technol.* **14**: 163. **[119a]**

Martínez, F., and Labuza, T. P. (1968). Rate of deterioration of freeze-dried salmon as a function of relative humidity. *J. Food Sci.* **33**: 241. **[120]**

Mazza, G., and LeMaguer, M. (1978). Water sorption properties of yellow globe onion (*Allium cepa* L.). *Can. Inst. Food Sci. Technol. J.* **11**: 189. **[121]**

Mizrahi, S., and Karel, M. (1977). Moisture transfer in a packaged product in isothermal storage: extrapolating data to any package-humidity combination and evaluating water sorption isotherms. *J. Food Processing Preservation* **1**: 225. **[125]**

Mizrahi, S., Labuza, T. P., and Karel, M. (1970). Computer-aided predictions of extent of browning in dehydrated cabbage. *J. Food Sci.* **35**: 799. **[124]**

Mossel, D. A. A. (1975). Water and micro-organisms in foods—A synthesis. *In* "Water Relations of Foods" (R. B. Duckworth, ed.), p. 347. Academic Press, New York. **[126]**

Multon, J. L., and Guilbot, A. (1975). Water activity in relation to the thermal inactivation of enzymic proteins. *In* "Water Relations of Foods" (R. B. Duckworth, ed.), p. 379. Academic Press, New York. **[127]**

Nemitz, G. (1961). Über die wasserbindung durch eiweisstoffe und deren verhalten während der trocknung, Dissertation; Technische Hochschule Karlsruhe. **[128]**

Ngoddy, P. O., and Bakker-Arkema, F. W. (1970). A generalized theory of sorption phenomena in biological materials (Part 1. The isotherm equation). *Trans. ASAE* **13**: 612. **[129]**

Nip, W. K. (1978). Development and storage stability of drum-dried guava and papaya-taro flakes. *J. Food Sci.* **44**: 222. **[130]**

Okwelogu, T. N., and Mackay, P. J. (1969). Cashewnut moisture relations. *J. Sci. Food Agric.* **20**: 697. **[131]**

Oswin, C. R. (1946). *J. Chem. Ind.* (*London*) **65**: 419. **[132]**

Palnitkar, M. P., and Heldman, D. R. (1971). Equilibrium moisture characteristics of freeze-dried beef components. *J. Food Sci.* **36**: 1015. **[133]**

Peleg, M., and Mannheim, C. H. (1977). The mechanism of caking of powdered onion. *J. Food Processing Preservation* **1**: 3. **[134]**

Peri, C., and De Cesari, L. (1974). Thermodynamics of water sorption on *Sacc. cerevisiae* and cell viability during spray-drying. *Lebensm. Wiss. Technol.* **7**: 56. **[135]**

Piñaga, F., and Lafuente, G. (1965). Horchata en polvo. I. Humedades de equilibrio de la horchata liofilizada. *Rev. Agroq. Tecnol. Alimentos* **5**: 99. **[136]**

Pitt, J. I. (1975). Xerophilic fungi and the spoilage of foods of plant origin. *In* "Water Relations of Foods" (R. B. Duckworth, ed.), p. 273. Academic Press, New York. **[137]**

Pixton, S. W., and Henderson, S. (1979). Moisture relations of dried peas, shelled almonds and lupins. *J. Stored Prod. Res.* **15**: 59. **[138]**

Pixton, S. W., and Warburton, S. (1971a). Moisture content/relative humidity equilibrium of some cereal grains at different temperatures. *J. Stored Prod. Res.* **6**: 283. **[139]**

Pixton, S. W., and Warburton, S. (1971b). Moisture content relative humidity equilibrium, at different temperatures, of some oilseeds of economic importance. *J. Stored Prod. Res.* **7**: 261. **[140]**

Pixton, S. W., and Warburton, S. (1973a). Determination of moisture content and equilibrium relative humidity of dried fruit—Sultanas. *J. Stored Prod. Res.* **8**: 263. **[141]**

Pixton, S. W., and Warburton, S. (1973b). The moisture content/equilibrium relative humidity relationship of macaroni. *J. Stored Prod. Res.* **9**: 247. **[142]**

Pixton, S. W., and Warburton, S. (1975a). The moisture content/equilibrium relative humidity relationship of soya meal. *J. Stored Prod. Res.* **11**: 249. **[143]**

Pixton, S. W., and Warburton, S. (1975b). The moisture content–equilibrium relative humidity relationship of rice bran at different temperatures. *J. Stored Prod. Res.* **11**: 1. **[144]**

Pixton, S. W., and Warburton, S. (1976). The relationship between moisture content and equilibrium relative humidity of dried figs. *J. Stored Prod. Res.* **12**: 87. [145]

Pixton, S. W., and Warburton, S. (1977a). The moisture content/equilibrium relative humidity relationship and oil composition of rapeseed. *J. Stored Prod. Res.* **13**: 77. [146]

Pixton, S. W., and Warburton, S. (1977b). The moisture content/equilibrium relative humidity relationship of a dried yeast product. *J. Stored Prod. Res.* **13**: 35. [147]

Potthast, K., Hamm, R., and Acker, L. (1975). Enzymic reactions in low moisture foods. In "Water Relations of Foods" (R. B. Duckworth, ed.), p. 365, Academic Press, New York. [148]

Potthast, K., Acker, L., and Hamm, R. (1977a). Einfluss der Wasseraktivität auf enzymatische Veränderungen in gefriergetrockneten Muskelfleisch. III. Abbau von Muskellipiden. *Z. Lebensm. Unters. Forsch.* **165**: 15. [149]

Potthast, K., Acker, L., and Hamm, R. (1977b). Einfluss der Wasseraktivität auf enzymatische Veränderungen in gefriergetrockneten Muskelfleisch. IV. Anderung der Aktivität glykolytischer Enzyme während der Lagerung. *Z. Lebensm. Unters. Forsch.* **165**: 18. [150]

Putranon, R., Bowrey, R. G., and Eccleston, J. (1979). Sorption isotherms for two cultivars of paddy rice grown in Australia. *Food Technol. Australia,* Dec. 1979, p. 510. [150a]

Quast, D. G., and Teixeira Neto, R. O. (1976). Moisture problems of foods in tropical climates. *Food Technol.* **30**: 98. [152]

Quast, D. G., Karel, M., and Rand, W. M. (1972). Development of a mathematical model for oxidation of potato chips as a function of oxygen pressure, extent of oxidation and equilibrium relative humidity. *J. Food Sci.* **37**: 673. [151]

Rasekh, J. G., Stilling, B. R., and Dubrow, D. L. (1971). Moisture adsorption of fish protein concentrate at various relative humidities and temperatures. *J. Food Sci.* **36**: 705. [153]

Resnik, S., and Chirife, J. (1979). Effect of moisture content and temperature on some aspects of nonenzymatic browning in dehydrated apple. *J. Food Sci.* **44**: 601. [154]

Rockland, L. B. (1957). A new treatment of hygroscopic equilibria: Application to walnuts (*Juglans regia*) and other foods. *Food Res.* **22**: 604. [155]

Rockland, L. B., and Nishi, S. K. (1980). Influence of water activity on food product stability. *Food Technol.* **34**: 42. [156]

Rodríguez Arias, J. H. (1956). Desorption isotherms and drying rates of shelled corn in the temperature range of 40 to 140°F. Ph.D. Thesis, Dept. of Agricultural Engineering, Michigan State University, East Lansing. [157]

Rüegg, M., and Blanc, B. (1976). Effect of pH on water vapor sorption by caseins. *J. Dairy Sci.* **59**: 109. [159]

Salwin, H. (1962). The role of moisture in deteriorative reactions of dehydrated foods. In "Freeze Drying of Foods" (F. R. Fisher, ed.), National Academy of Sciences–National Research Council, Washington, D.C. [160]

Salwin, H., and Slawson, V. (1959). Moisture transfer in combinations of dehydrated foods. *Food Technol.* **13**: 715. [161]

Sanjeevi, R., and Ramanathan, N. (1976). Moisture sorption hysteresis on raw and treated collagen fibers. *Indian J. Biochem. Biophys.* **13**: 98. [162]

San José, C., Asp, N. G., Burvall, A., and Dahlqvist, A. (1977). Water sorption in lactose hydrolyzed dry milk. *J. Dairy Sci.* **60**: 1539. [163]

Saravacos, G. D. (1967). Effect of the drying method on the water sorption of dehydrated apple and potato. *J. Food Sci.* **32**: 81. [164]

Saravacos, G. D. (1969). Sorption and diffusion of water in dry soybeans. *Food Technol.* **23**: 145. [165]

Saravacos, G. D., and Stinchfield, R. M. (1965). Effect of temperature and pressure on the sorption of water vapor by freeze-dried food materials. *J. Food Sci.* **30**: 779. [166]

Schwartz, T. A. (1943). Improvement needed in technique for testing food packages. *Food Ind.* **15**: 68. [167]

Scott, W. J. (1957). Water relations of food spoilage microorganisms. *Adv. Food Res.* **7**: 83. [168]

Shibata, S., Toyoshima, H., Imai, T., and Inoue, Y. (1976). Studies on storage of dried Japanese noodle. Part I. Relation between NaCl content of dried noodle (udon) and its equilibrium moisture. *Nippon Shokuhin Kogyo Gakkaishi* **23**: 397. **[169]**

Shotton, E., and Harb, N. (1965). The effect of humidity and temperature on the equilibrium moisture content of powders. *J. Pharm. Pharmacol.* **17**: 504. **[170]**

Smith, S. E. (1947). The sorption of water vapor by high polymers. *J. Am. Chem. Soc.* **69**: 646. **[171]**

Strolle, E. O., and Cording, J., Jr. (1965). Moisture equilibria of dehydrated mashed potato flakes. *Food Technol.* **19**: 171. **[173]**

Strolle, E. O., Cording, J., Jr., Mc Dowell, P. E., and Eskew, R. K. (1970). Effect of sucrose on crispness of explosion-puffed apple pieces exposed to high humidities. *J. Food Sci.* **35**: 338. **[172]**

Taylor, A. A. (1961). Determination of moisture equilibria in dehydrated foods. *Food Technol.* **15**: 536. **[174]**

Troller, J., and Christian, J. H. B. (1978). "Water activity and Food." Academic Press, New York. **[175]**

Vaidya, P. S., Verma, K. K., Rustagi, K. N., Jaisani, J. C., and Mathew, T. V. (1977). Studies on the equilibrium relative humidity and seasonal variation in moisture content of walnuts (*Juglans regia*). *J. Food Sci. Technol.* **14**: 169. **[176]**

van den Berg, C., and Bruin, S. (1978). Water activity and its estimation in food systems: theoretical aspects. *2nd Int. Symp. Properties of water in Relation to Food Quality and Stability (Isopow-II), September 10–16, Osaka, Japan*. **[178]**

van den Berg, C., and Leniger, X. (1976). The water activity of foods. *Proc. Int. Congr. Eng. Food, 9–13 August*, Boston. **[177]**

van Twisk, P. (1969). The sorption isotherms of maize meal. *J. Food Technol.* **4**: 75. **[179]**

Varshney, N. N., and Ojha, T. P. (1977). Water vapour sorption properties of dried milk baby foods. *J. Dairy Res.* **44**: 93. **[180]**

Viollaz, P. E., Iglesias, H. A., and Chirife, J. (1978). Slopes of moisture sorption isotherms of foods as a function of moisture content. *J. Food Sci.* **43**: 606. **[181]**

Volman, D. H., Simons, J. W., Seed, J. R., and Sterling, C. (1960). Sorption of water vapour by starch. Thermodynamics and structural changes for dextrin, amylose, and amylopectin. *J. Polymer Sci.* **46**: 355. **[182]**

von Roth, D. (1977). Das Wasserdampfsorptionsverhalten von Puderzucker. *Zucker* **30**: 274. **[158]**

Walker, J. E., Jr., Wolf, M., and Kapsalis, J. G. (1973). Adsorption of water vapor on myosin A and myosin B. *J. Agric. Food Chem.* **21**: 878. **[183]**

Warburton, S., and Pixton, S. W. (1973). Moisture relations of freshly harvested barley in store. *J. Stored Products Res.* **9**: 269. **[184]**

Warburton, S., and Pixton, S. W. (1975). The effect of the addition of glycerol on the moisture content/equilibrium relative humidity relationship of wheatfeed. *J. Stored Products Res.* **11**: 107. **[185]**

Weston, W. J., and Morris, H. J. (1954). Hygroscopic equilibria of dry beans. *Food Technol.* **8**: 353. **[186]**

White, G. W., and Cakebread, S. H. (1966). The glassy state in certain sugar-containing food products. *J. Food Technol.* **1**: 73. **[187]**

Wink, W. A., and Sears, G. R. (1950). Instrumentation studies. LVII. Equilibrium relative humidities above saturated salt solutions at various temperatures. *TAPPI* **33**: 96A. **[188]**

Wolf, M., Walker, J. E., Jr., and Kapsalis, J. G. (1972). Water vapor sorption hysteresis in dehydrated foods. *J. Agric. Food Chem.* **20**: 1073. **[189]**

Wolf, W., Spiess, W. E. L., and Jung, G. (1973). Die Wasserdampfsorptionsisothermen einiger in der Literatur bislangwening berücksichtigter lebensmittel. *Lebensm. Wiss. Technol.* **6**: 94. **[190]**

Young, J. H. (1976). Evaluation of models to describe sorption and desorption equilibrium moisture content isotherms of Virginia-type peanuts. *Trans. ASAE* **19**: 146. **[191]**

PRODUCT INDEX*

Actomyosin, beef (1), 11
Agar-agar (2), 11
Albumin, egg (3) (5), 12, 13
Albumin, egg, coagulated (4), 12
Albumin serum (6), 13
Alginic acid (7) (8), 14, 14
Almonds (9) (10) (11) (12) (13), 15, 15, 16,
  16, 17
Amylopectin (14), 17
Amylose (15), 18
Anise (16) (17), 18, 19
Apple (18) (19) (20) (21) (22), 19, 20, 20, 21,
  21
Apple juice (23), 22
Apple, osmotically treated (24) (25) (26), 22,
  23, 23
Apricots (27) (28), 24, 24
Asparagus (29), 25
Avicel + safflower protein (376), 198
Avicel + starch gel (413), 217
Avocado (30) (31) (32), 25, 26, 26
Banana (33) (34) (35), 27, 27, 28
Barley (36) (37) (38), 28, 29, 29
Beans (39) (40) (41), 30, 30, 31
Beef (42) (43) (44) (45) (46) (47) (48) (49),
  31, 32, 32, 33, 33, 34, 34, 35
Beet root (50) (51) (52), 35, 36, 36
Beet root juice (53), 37
Beet root sugar (54), 37
Beet root sugar, water insoluble components
  (55), 38
Blackcurrants (56), 38
Blackcurrant juice (178), 99
Borsch, instant (57), 39
Broccoli (58), 39
Cabbage (59) (60) (61) (62) (63), 40, 40, 41,
  41, 42

Cake (64) 42
Cardamon (65) (66), 43, 43
β-Carotene-cellulose model (67), 44
Carrots (68) (69) (70) (71) (72) (73) (74) (75)
  (76) (77), 44, 45, 45, 46, 46, 47, 47, 48,
  48, 49
Carrot seeds (78), 49
Casein (79) (80) (81) (100) (243), 50, 50, 51,
  60, 132
Caseinate sodium (82), 51
Cashewnuts (83) (84), 52, 52
Celery (85) (86), 53, 53
Cell concentrate, red blood (87), 54
Cellulose, microcrystalline (88), 54
Cellulose microcrystalline-oil model (89) (90)
  (91) (92), 55, 55, 56, 56
Carboxymethyl cellulose, sodium (93), 57
Chamomile tea (94) (95), 57, 58
Cheese, Edam (96) (97), 58, 59
Cheese, Emmental (98) (99), 59, 60
Cheese (100), 60
Cottage cheese whey (101) (102), 61, 61
Chicken (103) (104) (105) (106) (107), 62, 62,
  63, 63, 64
Chives (108), 64
Cinnamon (109) (110), 65, 65
Citric acid (111), 66
Citrus juice (112), 66
Cloves (113) (114), 67, 67
Cocoa (115), 68
Cod (116), 68
Coffee "Inka" (117), 69
Coffee beans (118) (119) (120), 69, 70, 70
Coffee, roasted and ground (121) (123), 71,
  72
Coffee, roasted and ground, decaffeinated
  (122), 71

*Numbers in parentheses refer to figure number(s) in Chapter 1, Section 2.

Coffee extract (124) (126), 72, 73
Coffee extract, decaffeinated (125), 73
Coffee extract, decaffeinated, agglomerated (127), 74
Coffee extract, agglomerated (128), 74
Coffee products (129), 75
Collagen (130) (131), 75, 76
Collagen, chrome-tanned (132), 76
Collagen, enzyme treated (133), 77
Collagen, myrab-tanned (134), 77
Connective tissue, beef (135), 78
Copra, smoked (136), 78
Coriander (137) (138), 79, 79
Corn (139) (140) (141) (149) (150) (151), 80, 80, 81, 85, 85, 86
Corn flour (142) (144) (146) (147) (148), 81, 82, 83, 84, 84
Corn flour, degermed (143), 82
Corn germ flour (145), 83
Corn-soy flour (142), 81
Cottonseed meal (152), 86
Curd gel (153) (154), 87, 87
Dextran-10 (155), 88
Dextrin (156) (157), 88, 89
Eggs (158) (159) (160) (164), 89, 90, 90, 92
Egg white (161) (162) (163), 91, 91, 92
Egg yolk (162) (165), 91, 93
Eggplant (166), 93
Elastin (167), 94
Fennel tea (168) (169), 94, 95
Figs (170), 95
Fish proteins, myofibrillar, enzimatically modified (171) (172) (173) (174), 96, 96, 97, 97
Fish protein concentrate (175) (176), 98, 98
Fructose (177) (429), 99, 225
Fruit juices (178), 99
Gelatin (179) (180) (181), 100, 100, 101
Gelatin-microcrystalline cellulose (182), 101
Gelatin-starch gel (183), 102
Ginger (184) (185), 102, 103
Glucose (186) (429), 103, 225
Glucose-starch gel (414), 217
Glutamate, sodium (187), 104
Gluten, wheat (188) (189) (190), 104, 105, 105
Grapefruit (191) (192), 106, 106
Grass seed (193), 107
Green peas (301) (302), 161, 161
Green pepper (194), 107
Groundnuts (195) (196) (197), 108, 108, 109
Guava-taro (198), 109
Halibut (199), 110

Hazelnut kernels (200), 110
Hibiscus tea (201) (202), 111, 111
Hops (203), 112
Horseradish (204) (205), 112, 113
Icing sugar (431) (432) (433) (434) (435) (436), 226, 226, 227, 227, 228, 228
β-Lactoglobulin (206) (207), 113, 114
Lactose (208) (209) (210) (211), 114, 115, 115, 116
Laurel (212) (213), 116, 117
Leek (214) (215), 117, 118
Lemon crystals (216), 118
Lentil (217) (218), 119, 119
Linseed seed (219) (220), 120, 120
Lupins, bitter blue (221) (222), 121, 121
Macaroni (223) (224) (225), 122, 122, 123
Maltose (226) (227), 123, 124
Marrow (228), 124
Milk, baby food (229) (230), 125, 125
Milk (231) (232) (233) (234) (235) (236) (237) (244) (245) (246) (247) (248) (249) (250), 126, 126, 127, 127, 128, 128, 129, 132, 133, 133, 134, 134, 135, 135,
Milk, lactose hydrolyzed (238) (239) (240) (241), 129, 130, 130, 131
Milk-orange, drink (242), 131
Milk components (243), 132
Muscle (251) (252), 136, 136
Mushrooms (253) (254) (255) (256) (257) (258), 137, 137, 138, 138, 139, 139
Mushrooms, liquid extract (259), 140
Myosin A (260), 140
Myosin B (261) (262), 141, 141
Noodles (263) (264) (265) (266) (267) (268) (269), 142, 142, 143, 143, 144, 144, 145
Nutmeg (270) (271), 145, 146
Peanut oil (272), 146
Oleic acid (273), 147
Onion (274) (275) (276) (277) (278), 147, 148, 148, 149, 149
Onion seed (279), 150
Orange crystals (178) (280) (281), 99, 150, 151
Orange-milk drink (242), 131
Orgeat (282), 151
Ovalbumin (283), 152
Papaya-taro (284), 152
Paranut (285) (286), 153, 153
Parsley (287), 154
Parsnip seed (288), 154
Pea flour (289), 155
Peach (290) (291), 155, 156

Peanuts (292) (293) (294) (296), 156, 157, 158
Peanut shell (295), 158
Pear (297) (298), 159, 159
Peas (299) (300), 160, 160
Green peas (301) (302), 161, 161
Peas seed (303), 162
Pekanut (304) (305), 162, 163
Peppermint tea (306) (307), 163, 164
Pimiento (308), 164
Pineapple (309), 165
Pistachio nut (310) (311) (312) (313), 165, 166, 166, 167
Plum (314), 167
Plum juice (315), 168
Pork (316) (317), 168, 169
Potato (318) (319) (320) (321) (322) (323) (324) (325) (326) (327) (328), 169, 170, 170, 171, 171, 172, 172, 173, 173, 174, 174
Potato flakes (329) (330), 175, 175
Prunes (331) (332), 176, 176
Pseudoglobulin, α, β, γ (333) (334), 177, 177

Radish (335), 178
Raisins (336), 178
Rapeseed (337) (338) (339) (340) (341) (342) (343) (344), 179, 179, 180, 180, 181, 181, 182, 182
Raspberry (345), 183
Rhubarb (346), 183
Rice (347) (348) (349) (350) (351) (352) (353) (354) (355) (356) (357) (358) (359) (360) (361) (362) (363) (364) (365), 184, 184, 185, 185, 186, 186, 187, 187, 188, 188, 189, 189, 190, 190, 191, 191, 192, 192, 193
Rice bran, unextracted (366) (367) (372) (373), 193, 194, 196, 197
Rice bran, extracted (368) (369), 194, 195
Rice bran, unextracted, parboiled (370) (371), 195, 196

Sacc. cerevisiae (374), 197
Safflower protein (375), 198
Safflower protein + avicel (376), 198
Safflower protein + starch (377), 199
Safflower protein + starch gel (378), 199
Salmin (379), 200
Salmon (380), 200
Salsifi (381) (382), 201, 201
Sarcoplasmic fraction, beef (383), 202
Sorghum (384) (385) (386) (387) (388), 202, 203, 203, 204, 204,

Soup (389) (390) (391) (392), 205, 205, 206, 206
Soybeans (393) (394) (395) (396) (399) (400) (401), 207, 207, 208, 208, 210, 210, 211
Soy flour (142), 81
Soy-corn flour (142), 81
Soya meal (397) (398), 209, 209
Soy protein concentrate (402) (403), 211, 212
Soy protein isolate (404) (405), 212, 213
Soy-whey beverage (483) (490), 252, 255
Spinach (406) (407), 213, 214
Starch (408) (409) (410) (411) (412) (415) (416) (417) (418) (419) (420) (421) (422) (423), 214, 215, 215, 216, 216, 218, 218, 219, 219, 220, 220, 221, 221, 222
Starch + safflower protein (377), 199
Starch gel + avicel (413), 217
Starch gel + safflower protein (378), 199
Starch-glucose gel (414), 217
Strawberry (424) (425), 222, 223
Sucrose (426) (427) (428) (429), 223, 224, 224, 225
Sugars (429), 225
Sugar beet root (54), 37
Sugar beet root, water insoluble components (55), 38
Sugar cane products (430), 225
Sugar icing (431) (432) (433) (434) (435) (436), 226, 226, 227, 227, 228, 228
Sultanas (437) (438), 229, 229
Sunflower seed (439), 230
Sweet marjoram (440) (441), 230, 231
Tapioca (442) (443), 231, 232
Taro-guava (198), 109
Taro-papaya (284), 152
Tea, black (444) (445) (446) (447) (450), 232, 233, 233, 234, 235
Tea, fannings (448), 234
Tea, infusion (449), 235
Thyme (451), (452), 236, 236
Tomato (453) (454), 237, 237
Trout (455) (456) (457) (458), 238, 238, 239, 239
Tuna, Big Eye (459), 240,
Turkey (460) (461), 240, 241
Waffer sheet (462), 241
Walnut kernels (463) (464) (465) (466) (467), 242, 242, 243, 243, 244
Wheat (468) (469) (470) (471) (472) (473) (474) (475) (476) (477), 244, 245, 245, 246, 246, 247, 247, 248, 248, 249

Wheatfeed   (478) (479) (480), 249, 250, 250
Wheat flour   (481) (482), 251, 251
Whey-soy beverage   (483) (490), 252, 255
Whey   (243) (484) (489), 132, 252, 255
Whey protein concentrate   (485) (486) (487)
    (488), 253, 253, 254, 254

Winter savory   (491) (492), 256, 256
Yams   (493), 257
Yeast   (374) (494) (495) (496) (497) (498)
    (499), 197, 257, 258, 258, 259, 259, 260
Yoghurt   (500) (501), 260, 261

FOOD SCIENCE AND TECHNOLOGY

A SERIES OF MONOGRAPHS

Maynard A. Amerine, Rose Marie Pangborn, and Edward B. Roessler, PRINCIPLES OF SENSORY EVALUATION OF FOOD. 1965.

S. M. Herschdoerfer, QUALITY CONTROL IN THE FOOD INDUSTRY. Volume I — 1967. Volume II — 1968. Volume III — 1972.

Hans Reimann, FOOD-BORNE INFECTIONS AND INTOXICATIONS. 1969.

Irvin E. Leiner, TOXIC CONSTITUENTS OF PLANT FOODSTUFFS. 1969.

Martin Glicksman, GUM TECHNOLOGY IN THE FOOD INDUSTRY. 1970.

L. A. Goldblatt, AFLATOXIN. 1970.

Maynard A. Joslyn, METHODS IN FOOD ANALYSIS, second edition. 1970.

A. C. Hulme (ed.), THE BIOCHEMISTRY OF FRUITS AND THEIR PRODUCTS. Volume 1 — 1970. Volume 2 — 1971.

G. Ohloff and A. F. Thomas, GUSTATION AND OLFACTION. 1971.

George F. Stewart and Maynard A. Amerine, INTRODUCTION TO FOOD SCIENCE AND TECHNOLOGY. 1973.

C. R. Stumbo, THERMOBACTERIOLOGY IN FOOD PROCESSING, second edition. 1973.

Irvin E. Liener (ed.), TOXIC CONSTITUENTS OF ANIMAL FOODSTUFFS. 1974.

Aaron M. Altschul (ed.), NEW PROTEIN FOODS: Volume 1, TECHNOLOGY, PART A — 1974. Volume 2, TECHNOLOGY, PART B — 1976. Volume 3, ANIMAL PROTEIN SUPPLIES, PART A — 1978. Volume 4, ANIMAL PROTEIN SUPPLIES, PART B — 1981.

S. A. Goldblith, L. Rey, and W. W. Rothmayr, FREEZE DRYING AND ADVANCED FOOD TECHNOLOGY. 1975.

R. B. Duckworth (ed.), WATER RELATIONS OF FOOD. 1975.

Gerald Reed (ed.), ENZYMES IN FOOD PROCESSING, second edition. 1975.

A. G. Ward and A. Courts (eds.), THE SCIENCE AND TECHNOLOGY OF GELATIN. 1976.

John A. Troller and J. H. B. Christian, WATER ACTIVITY AND FOOD. 1978.

A. E. Bender, FOOD PROCESSING AND NUTRITION. 1978.

D. R. Osborne and P. Voogt, THE ANALYSIS OF NUTRIENTS IN FOODS. 1978.

Marcel Loncin and R. L. Merson, FOOD ENGINEERING: PRINCIPLES AND SELECTED APPLICATIONS. 1979.

Hans Riemann and Frank L. Bryan (eds.), FOOD-BORNE INFECTIONS AND INTOXICATIONS, second edition. 1979.

N. A. Michael Eskin, PLANT PIGMENTS, FLAVORS AND TEXTURES: THE CHEMISTRY AND BIOCHEMISTRY OF SELECTED COMPOUNDS. 1979.

J. G. Vaughan (ed.), FOOD MICROSCOPY. 1979.

J. R. A. Pollock (ed.), BREWING SCIENCE, Volume 1 — 1979. Volume 2 — 1980.

Irvin E. Liener (ed.), TOXIC CONSTITUENTS OF PLANT FOODSTUFFS, second edition. 1980.

J. Christopher Bauernfeind (ed.), CAROTENOIDS AS COLORANTS AND VITAMIN A PRECURSORS: TECHNOLOGICAL AND NUTRITIONAL APPLICATIONS. 1981.

Pericles Markakis (ed.), ANTHOCYANINS AS FOOD COLORS. 1982.

Vernal S. Packard, HUMAN MILK AND INFANT FORMULA. 1982.

George F. Stewart and Maynard A. Amerine, INTRODUCTION TO FOOD SCIENCE AND TECHNOLOGY, SECOND EDITION. 1982.

Malcolm C. Bourne, Food Texture and Viscosity: Concept and Measurement. 1982.

R. Macrae (ed.), HPLC in Food Analysis. 1982.

Héctor A. Iglesias and Jorge Chirife, Handbook of Food Isotherms: Water Sorption Parameters for Food and Food Components. 1982.

In preparation

John A. Troller, Sanitation in Food Processing. 1983.